W9-COI-824

Series/Number 07–005

COHORT ANALYSIS

NORVAL D. GLENN
University of Texas

SAGE PUBLICATIONS
International Educational and Professional Publisher
Thousand Oaks   London   New Delhi

For information:

Sage Publications, Inc.
2455 Teller Road
Thousand Oaks, California 91320
E-mail: order@sagepub.com

Sage Publications Ltd.
1 Oliver's Yard
55 City Road
London, EC1Y 1SP
United Kingdom

Sage Publications India Pvt. Ltd.
B-42, Panchsheel Enclave
Post Box 4109
New Delhi 110 017 India

Printed in the United States of America on acid-free paper.

Library of Congress Cataloging-in-Publication Data

Glenn, Norval D.
Cohort analysis / Norval D. Glenn.—2nd ed.
    p. cm. — (Quantitative applications in the social sciences ; 07-005)
Includes bibliographical references and index.
ISBN 0-7619-2215-6 (pbk.)
    1. Cohort analysis. I. Title. II. Series: Sage university papers series. Quantitative applications in the social sciences ; 07-005, 2nd ed.
HB849.47.G584 2005
001.4'2—dc22

                                        2004019523

05   06   07   08   09   10   9   8   7   6   5   4   3   2   1

Acquiring Editor:	Lisa Cuevas Shaw
Editorial Assistant:	Margo Crouppen
Project Editor:	Claudia A. Hoffman
Copy Editor:	Liann Lech
Typesetter:	C&M Digitals (P) Ltd.
Indexer:	Molly Hall

CONTENTS

SERIES EDITOR'S INTRODUCTION

The concept of *cohort* has been central in many disciplines of the social sciences, such as demography, epidemiology, psychology, political science, and sociology. An example of the concept in a major work is Glen Elder's *Children of the Great Depression: Social Change in Life Experience* first published in 1974 by the University of Chicago Press. The children in this book grew up in difficult times, and they had to fight for things that people take for granted today. The individuals in this cohort, initially thought of as a "lost generation," came back from the war to prove themselves as the can-do generation. It was their unique experience that helped forge the traits of the cohort.

Broadly speaking, cohort analysis is quantitative research using a measure of the concept of *cohort* and studying its effect on some outcome variable(s), such as the postwar behavior of the Great Depression cohort. In this sense, a cohort consists of people who share a common experience during a specified period of time. Most often, the term *cohort* alone refers to a human birth cohort. While sharing a birth year (or a group of multiple years) can definitely expose individuals in a cohort to similar social changes, people of different ages can also belong to the same cohort, such as graduate students who started their graduate training in a given year.

Since the early 1970s when Elder's *Children of the Great Depression* was published, researchers have demonstrated intense interest in how to uniquely estimate the statistical cohort effect of a particular social behavior or phenomenon. This is no small task because, in addition to estimating the effect of cohort (year of birth), analysts must estimate the effects of age (year since birth) and period (current year) at the same time. As is well known, without identifying restrictions, linear and additive Age, Period, Cohort (or APC) models are not identifiable because APC are exact linear functions of each other because of the identity $A = P - C$.

Since Norval Glenn's *Cohort Analysis* first came out in 1977, new methods that deal with this identification problem have been developed. We are excited about the second edition, which is entirely newly written, as it presents an updated account of cohort analysis covering both the methodological basics—including data issues—and a detailed discussion of some of the major alternative strategies for estimating age, period, and cohort effects. Social science researchers in a wide range of disciplines, especially those studying aging as well as social and cultural change, will benefit from this second edition.

—*Tim Futing Liao*
Series Editor

P.S. Let me thank my predecessor, Michael Lewis-Beck, who did all the actual editing of this title. I get only the honor of writing the introduction.

PREFACE

The first edition of *Cohort Analysis* was written almost 30 years ago, when the cohort approach to studying social and psychological phenomena was in its infancy. Although most of the material in that publication is still pertinent, this edition is completely new, lacking even a single paragraph lifted from the earlier one. I decided on a rewrite rather than a revision for several reasons. For instance, I have changed my mind about some matters, new analysis strategies have been introduced, and the treatment of data sources in the first edition is badly dated. In common with several of the early contributions to the Quantitative Applications in the Social Sciences Series, the 1977 edition of *Cohort Analysis* was written to be understandable to, among others, undergraduate students with only rudimentary knowledge of methods and statistics. In order to bring the level of sophistication in this edition closer to that in most of the later contributions to the series, I have assumed that the reader understands the basics of statistics up through multiple regression and is familiar with the core issues covered in a good upper-division undergraduate social science methods course. However, I have tried to keep *Cohort Analysis* one of the most understandable of the QASS papers. The first edition was mildly cautionary; this one is more strongly so. Numerous authors have cited the first edition while ignoring the cautions in it, occasionally even citing it as the authority for inappropriate applications of cohort analysis. A fairly typical citation has been after a statement that separating age, period, and cohort effects is difficult. I hope that authors who cite this edition will recognize that, except under conditions that hardly ever exist, a definitive separation of age, period, and cohort effects is not just difficult, but impossible. However, I also hope that they will realize that a definitive separation of the effects is not necessary in order for cohort analysis to be useful.

COHORT ANALYSIS, SECOND EDITION

NORVAL D. GLENN
University of Texas at Austin

1. INTRODUCTION

Cohort analysis, which is a general strategy for examining data rather than a specific statistical technique, has become increasingly popular in the social sciences in the past few decades as an abundance of data appropriate for its application has become available. Its popularity results partly from its usefulness in addressing substantive issues relating to aging and social and cultural change, but many researchers apparently have become interested in cohort analysis primarily because it presents an unusually intriguing methodological challenge. Predictably, this kind of interest has not been entirely salutary. Many cohort studies have been little more than exercises, and much effort has been expended in attempts to do things that are logically impossible. Many overly confident and probably incorrect conclusions have been drawn from the research.

Therefore, the purpose of this monograph is to discourage mechanical, ill-conceived cohort studies, as well as to encourage appropriate ones. The potential contribution to knowledge of the cohort approach is great but will be realized only if there is increased caution and sophistication among those who use it.

The Purposes of Cohort Analysis

Cohort analysis has been used primarily for two distinct purposes, although both have motivated some studies.

The more common of these is to assess the effects of aging, usually of human beings but sometimes of entities such as marriages. Obviously, the ideal research design for assessing causation, the randomized experiment, cannot be used in studies of aging; it is impossible to divide randomly a number of subjects into an experimental and a control group, and to allow the former to age and to withhold the stimuli associated with aging from the control group. Therefore, students of aging are left with a number of non-experimental and quasi-experimental methods, all of which have serious

1

limitations but among which cohort analysis has some unique advantages. I compare some of the main methods in the next section of this chapter. The primary focus of this monograph is on studies of aging, although most of the discussion is pertinent to other uses of cohort analysis as well.

The second major use of cohort analysis is to understand the sources and nature of social, cultural, and political change. Studies with this purpose account for much of the growth in popularity of cohort analysis in recent years and may become more common than cohort studies of aging. The cohort approach by no means stands alone as a useful strategy for quantitative research on social and cultural change, but there are no widely used methods competing to provide the kind of insights that can be gained by cohort studies. Therefore, I provide no systematic comparison with other methods for studying change. Although, in this monograph, I emphasize the use of cohort analysis to study aging, I devote a chapter to the cohort approach to studying change.

Definitions; Comparison
of Cohort Analysis With Related Methods

The term *cohort* originally referred to a group of warriors or soldiers, and the term is now sometimes used in a very general sense to refer to a number of individuals who have some characteristic in common. Here and in other literature on cohort analysis, however, the term is used in a more restricted sense to refer to those individuals (human or otherwise) who experienced a particular event during a specified period of time. The kind of cohort most often studied by social scientists is the human *birth cohort,* that is, those persons born during a given year, decade, or other period of time. If the term cohort is used without an adjective, it is usually understood that the referent is a human birth cohort. Otherwise, the event commonly experienced by the individuals is used as an adjective to identify the kind of cohort, as in retirement cohort and graduation cohort, and if the individuals are not human beings, another adjective may be needed to identify them. The events that define cohorts may range from marriage to joining an organization, from entering a graduate program to becoming a parent for the first time. The individuals are usually human beings but may be marriages, organizations, textbooks, movies, automobile models, or other entities that came into being during a specified period of time.

An unfortunate term that sometimes appears in the social scientific literature is *age cohort,* which fails to follow the rule that each kind of cohort be identified by the event that defines it. Age is, of course, a condition, and a changing one, rather than an event. When the term age cohort is used, often it is not clear whether the referent is a birth cohort or an age category, and the two are distinctly different. For instance, persons born in 1960 are a birth cohort, and their age will be different depending on when they are studied. Persons 30 years old are an age category, the membership of which changes each year as one 1-year birth cohort moves out and another moves in.

The term *cohort analysis* is usually reserved for studies in which two or more cohorts are compared with regard to at least one dependent variable measured at two or more points in time. Examples of studies that would not ordinarily be considered cohort analyses, even though they involve cohorts in some way, are (a) a comparison of the political attitudes of young adults, middle-aged persons, and elderly persons with data from one survey conducted in November 2002, and (b) a comparison of the attitudes of all members of a high school graduating class of 1985 in that year and in 1995. The former study is *cross-sectional,* that is, it is conducted with data collected at one point in time, or, more accurately, within a short period of time. The latter is a *panel study,* that is, it measures the characteristics of the same individuals at more than one point in time. Both kinds of studies can be very useful for some purposes, but they have limited value for attainment of the most common goal of cohort analysis, namely, assessment of the consequences of growing older, or, in other words, estimation of *age effects.*

Differences by age shown by cross-sectional data may or may not be age effects, because the people (or other entities) of different ages are members of different cohorts and may have been shaped by different formative experiences and influences. In other words, differences among them may be *cohort effects.* To illustrate, persons born in the United States in 1920 spent their late adolescence and early adulthood in the Great Depression, whereas persons born just 10 years later spent the same stages of life in a period of unprecedented prosperity and economic growth. This contrast is believed to have created lifelong differences between the cohorts in economic and political attitudes and behavior. Furthermore, small birth cohorts—those born when the birth rate is low—have different economic, educational, occupational, and even marital opportunities than larger ones, and these opportunities affect attitudes and behavior. Therefore, the problem with cross-sectional research for estimating either age or cohort effects is that it confounds the two. For instance, one cannot tell whether differences between middle-aged and elderly persons are the result of changes the elderly persons experienced as (and because) they grew older or whether they are the result of differences in cohort-based influences.

Another limitation of cross-sectional studies for estimating age effects is that differences between older and younger surviving individuals may result from a correlation of the dependent variable with longevity. Consider, for instance, the fact that recent national surveys in the United States have shown that a larger percentage of middle-aged than of elderly men say they drink alcoholic beverages. This finding by itself is not strong evidence that men tend to stop drinking as they grow older. The difference could, of course, be at least partly a cohort effect, and it could also be a *compositional effect* due to a tendency for heavy drinkers to die at a younger age than moderate drinkers or abstainers. Most cohort analyses use data from a series of cross-sectional studies, and thus, they also may confound the effects of differential mortality with age and/or cohort effects.

Panel studies can provide evidence of compositional effects due to differential mortality, and they also have the advantage of being able to deal with individual-level change as well as aggregate-level change in aging cohorts. However, they alone cannot provide convincing evidence of age effects, because not all changes that occur in individuals as they grow older are the results of aging. Especially in modern societies, people grow older not in a static society but in a changing one, and influences from social and cultural change impinge on persons as they grow older, bringing about changes in attitudes, behavior, health, and emotional states or offsetting effects that would result from aging in a static society. These *period effects* are confounded with age effects in the data from panel studies. To illustrate, in the United States, the birth cohorts that were in young adulthood in the 1970s had, as a whole, become more conservative in several respects by the late 1980s. One cannot tell how much, if any, of this shift resulted from influences associated with aging; the fact that the society as a whole changed in the same direction as the cohorts suggests that much, if not most, of the intracohort trend was brought about by period influences. Therefore, panel studies that gauged the political attitudes of high school seniors in the mid-1970s and again 10 years later fail to provide strong evidence of any effects of the transition to adulthood.

Another limitation of panel studies for studying the effects of human aging is that the persons studied may be affected by their participation in the studies. If that effect is on a dependent variable used in the study, it is known as a *panel conditioning effect,* which is confounded in panel data with any age and period effects. To state the problem differently, if there are panel conditioning effects, a sample of study participants originally representative of a larger population will become less representative through the waves (repeated data collections) of the study.

Panel conditioning effects may occur in several different ways. For instance, if the dependent variable is an ability or skill measured by a standardized test, there may be a special kind of panel conditioning effect known as a *practice effect;* that is, persons may become better at test taking through the waves of the panel study. Similarly, being repeatedly asked probing questions on a topic may make respondents more reflective and thus may lead to a change in attitudes. Or, asking persons detailed questions

TABLE 1.1

Percentage of Women Who
Were Married, by Age and Year, United States

	Year			
Age	1968	1978	1988	1998
25–34	87.4	76.6	67.3	67.3
35–44	87.1	82.1	76.3	72.1
45–54	82.4	80.5	76.2	70.8
55–64	67.7	70.4	70.7	67.8
65–74	46.5	48.3	53.3	54.8

SOURCE: Data are from the March Current Population Survey conducted by the U.S. Census Bureau. The percentages are from, or are calculated from data in, U.S. Census Bureau (1969, Table 37; 1979, Table 51; 1990, Table 49; 1999; Table 63).

about their jobs, marriages, or other aspects of their lives may be a form of consciousness raising that will lead to changes in satisfaction with the objects of the questions. Panel studies designed to deal with adjustment to aging are particularly vulnerable to panel conditioning effects, because the persons studied may learn effective modes of adjustment from their participation in the research.

It is possible to do cohort analyses with panel data, but few such studies have been conducted, and the usual kind of cohort analysis, which uses data from two or more cross-sectional studies, avoids panel conditioning effects. With this design, called *repeated cross-sectional,* no individual is studied at more than one point in time. Rather, different samples of individuals from each cohort are studied at different times.

A heuristic device useful for explaining the logic of cohort analysis is the *standard cohort table,* which is constructed by juxtaposing sets of cross-sectional data that show the relationship between age and some dependent variable, with the age intervals equal to the intervals between periods for which there are data. A simple table of this sort, which reports U.S. Census data on the percentage of women who were married, is Table 1.1.

Each column in the table is a set of cross-sectional data, in which age, cohort, and, at least to some small degree, compositional effects are confounded. In each row, in which there are data on four different birth cohorts when they were at the same age level, period and cohort effects are confounded. Each cohort represented in the table, except the one that was ages 25–34 in 1988 and the one that was 65–74 in 1968, can be traced for at least 10 years as it grew older by starting in the leftmost cell in which it is represented and reading diagonally down and to the right. In the data in each of these *cohort diagonals,* age and period effects are confounded, as are at least small compositional effects in the case of the cohorts that can be traced beyond middle age.

The data in each column of the standard cohort table suffer from the same confounding of effects as the data from a cross-sectional study, and the data in each cohort diagonal confound age and period effects in the same way as do the data from a panel study. However, in the cohort table, there are multiple columns and multiple cohort diagonals, which has raised the hopes of many researchers for a way to use statistical procedures to separate age, period, and cohort effects. More than a quarter of a century ago, I (Glenn, 1976) called attempts to separate statistically the effects confounded in cohort data "a futile quest," which it is, except in the unlikely event that all effects are nonlinear. However, not everyone has abandoned the quest. The continued search for a statistical technique that can be mechanically applied always to correctly estimate the effects is one of the most bizarre instances in the history of science of repeated attempts to do something that is logically impossible.

The Identification Problem

The impossibility of statistically separating age, period, and cohort effects (except in rare circumstances) grows out of the *identification problem*, which exists whenever three or more independent variables need to be included in an analysis and each one is a linear function of the others. In other words, the multiple correlation of each independent variable with the other ones is unity—the most extreme kind of collinearity that is possible. If all of the independent variables are entered as predictor variables in a regression equation, or are used in any similar kind of analysis, the computer program will not run; with the other variables controlled, the variance of each independent variable is zero. The identification problem occurs, for instance, when there is reason to think that a difference between two variables may affect a dependent variable and when each of those two variables may also affect the dependent variable. An example is a study of marital quality in which the husband's characteristics, the wife's characteristics, and the husband-wife difference may all affect the quality. Another example is a study of the effects of social mobility on psychological well-being, in which stratum of origin, stratum of destination, and the difference between the two may all affect well-being.

In cohort analysis, the three interrelated variables are, of course, age, period, and cohort, each of which is a perfect linear function of the other two. That is, knowledge of an individual's value on two of the variables provides knowledge of the value on the third. For instance, if one knows that a survey respondent was interviewed in 1990 when she was 20 years old, one knows approximately when she was born, or her birth cohort membership. Or, from knowledge that a person born in 1970 was interviewed in 1990, one can deduce that the person was about 20 years old at the time of the interview. So, if period and cohort (or any two of the three variables) are already entered as predictor variables in a regression equation, entering age (or whatever variable is excluded) provides redundant information and prevents the program from running. If there is reason to believe that either age, period, or cohort has no effects, then the effects of the remaining two variables can be estimated easily, but there is no straightforward way to estimate simultaneously the effects of all three variables.

The nature of the identification problem is illustrated by the hypothetical data in Tables 1.2, 1.3, and 1.4, in which the simplest interpretation of the linear variation of the values of the dependent variable in each table is that it reflects only age effects (Table 1.2), period effects (Table 1.3), or cohort effects (Table 1.4). (For the purpose of this illustration, it is assumed that the dependent variable is uncorrelated with longevity and, thus, that there are no compositional effects due to differential mortality.) However, the data in each table are also amenable to a two-variable interpretation, as the alternative explanations at the bottom of each table indicate, and the pattern of variation in each table could also result from an infinite number of different combinations of age, period, *and* cohort effects. Obviously, no statistical technique, by itself, can select among the different combinations of effects that could produce the same data. The selection has to be made by the researcher on the basis of theory and what he or she knows about the phenomenon or phenomena being studied from sources other than the cohort data being analyzed. In other words, the selection has to be made on the basis of what Philip Converse (1976) has called *side information*. This information sometimes comes from additional variables (other than age, period, cohort, and the dependent variable) recorded in the data set used for the study, but usually it comes from other sources.

One might think that although either a one-, two-, or three-variable explanation for a linear pattern of variation of cohort data is logically possible, the one-variable explanation would always be the most plausible. However, that is not the case. Even when the pattern of variation is almost as simple as that in Tables 1.2, 1.3, and 1.4, a two- or three-variable explanation is sometimes most consistent with theory and what is known about the dependent variable from sources other than the cohort data. Furthermore, the variation, even when basically linear, is usually considerably more complex than that in the hypothetical data in the tables, and complex patterns suggest complex explanations.

Consider, for instance, the case of age and reported job satisfaction in the United States (see Table 1.5). The overall level of job satisfaction has been

TABLE 1.2

Pattern of Data Showing Pure Age Effects,
Offsetting Period and Cohort Effects, or a Combination of
Age Effects and Offsetting Period and Cohort Effects

				Year		
Age	1950	1960	1970	1980	1990	2000
20-29	50	50	50	50	50	50
30-39	55	55	55	55	55	55
40-49	60	60	60	60	60	60
50-59	65	65	65	65	65	65
60-69	70	70	70	70	70	70
70-79	75	75	75	75	75	75

NOTE: Numbers in the cells are hypothetical values of a dependent variable. Alternative explanations are that (a) each 10 years of aging produces a 5-point increase in the dependent variable, (b) there is a 5-point per 10 years positive period effect on the dependent variable and a 5-point per 10 years negative cohort effect, and (c) there is some combination of age and offsetting period and cohort effects on the dependent variable. An infinite number of combinations of such effects could produce the pattern of variation in the dependent variable shown in the table.

TABLE 1.3

Pattern of Data Showing Pure Period Effects,
Offsetting Age and Cohort Effects, or a Combination of
Period Effects and Offsetting Age and Cohort Effects

				Year		
Age	1950	1960	1970	1980	1990	2000
20-29	30	35	40	45	50	55
30-39	30	35	40	45	50	55
40-49	30	35	40	45	50	55
50-59	30	35	40	45	50	55
60-69	30	35	40	45	50	55
70-79	30	35	40	45	50	55

NOTE: Numbers in the cells are hypothetical values of a dependent variable. Alternative explanations are that (a) each passage of 10 years produces a 5-point increase in the dependent variable, (b) there is a 5-point per 10 years positive age effect on the dependent variable and a 5-point per 10 years positive cohort effect, and (c) there is some combination of period and offsetting age and cohort effects on the dependent variable. An infinite number of combinations of such effects could produce the pattern of variation in the dependent variable shown in the table.

rather stable over the past few decades, and there has been a largely stable positive relationship between age and reported satisfaction. The simplest explanation is that this stable relationship reflects only age effects or age effects combined with compositional effects due to movement into and out

TABLE 1.4

Pattern of Data Showing Pure Cohort Effects, Offsetting Age and Period Effects, or a Combination of Cohort Effects and Offsetting Age and Period Effects

	Year					
Age	1950	1960	1970	1980	1990	2000
20-29	50	55	60	65	70	75
30-39	45	50	55	60	65	70
40-49	40	45	50	55	60	65
50-59	35	40	45	50	55	60
60-69	30	35	40	45	50	55
70-79	25	30	35	40	45	50

NOTE: Numbers in the cells are hypothetical values of a dependent variable. Alternative explanations are that (a) each cohort that reaches adulthood is 5 points higher on the dependent variable than the cohort before it, (b) there is a 5-point per 10 years negative age effect on the dependent variable and a 5-point per 10 years positive period effect, and (c) there is some combination of cohort and offsetting age and period effects on the dependent variable. An infinite number of combinations of such effects could produce the pattern of variation in the dependent variable shown in the table.

TABLE 1.5

Percentage of Persons Who Said They Were *Very Satisfied*[a] With Their Work, by Year and Age, United States (*n* in parentheses)[b]

	Age					
Year	25-34	35-44	45-54	55-64	65-74	Total
1972-1976	44.4	53.9	51.8	60.1	63.4	52.4
	(1,475)	(1,208)	(1,358)	(936)	(313)	(5,290)
1977-1980	41.3	49.0	52.2	61.1	60.6	50.2
	(989)	(769)	(684)	(556)	(223)	(3,221)
1982-1986	42.4	49.1	54.1	57.2	62.4	52.9
	(1,868)	(1,384)	(1,106)	(865)	(320)	(5,543)
1987-1991	40.1	46.5	50.9	52.2	57.5	49.4
	(1,614)	(1,553)	(1,062)	(655)	(277)	(5,161)
1993-1996	39.3	44.6	47.3	55.9	68.1	45.9
	(1,483)	(1,563)	(1,246)	(584)	(183)	(5.059)
1998-2002	46.3	45.6	49.1	56.5	61.6	48.7
	(1,275)	(1,394)	(1,149)	(539)	(189)	(4,546)
Total	42.2	47.8	50.8	57.3	62.1	48.9
	(8,704)	(7,871)	(6,605)	(4,135)	(1,505)	(28,820)

SOURCE: The 1972-2002 United States General Social Surveys conducted by the National Opinion Research Center (Davis et al., 2002).

a. Other response alternatives are *moderately satisfied, a little dissatisfied,* and *very dissatisfied.*

b. Number of respondents is approximate due to weighting. Data are weighted by number of adults in the household divided by mean number of adults in GSS households.

TABLE 1.6

Pattern of Data Showing Nonlinear Variation in a
Dependent Variable That Can Reasonably Be Interpreted to
Reflect Only Age or Compositional Effects

| | Year | | | | | |
Age	1950	1960	1970	1980	1990	2000
20-29	50	50	50	50	50	50
30-39	52	52	52	52	52	52
40-49	62	62	62	62	62	62
50-59	62	62	62	62	62	62
60-69	50	50	50	50	50	50
70-79	45	45	45	45	45	45

NOTE: Numbers in the cells are hypothetical values of a dependent variable.

of the labor force. However, this stability in the overall level and age
pattern of job satisfaction seems curious in view of the known fact that, in
many respects, conditions of work have changed (probably on balance for
the better), and each successive cohort of young adults entering the labor
force in recent years apparently has had higher expectations for work conditions and rewards than the one before it. Therefore, it seems possible,
indeed likely, that the stable age pattern partly reflects offsetting period and
cohort effects. A departure from linearity in the relationship between age
and job satisfaction, whereby the increment of satisfaction associated with
a year of age is less in the middle adult years than at younger and older
ages, does suggest either age or compositional effects, making a three- or
four-variable explanation attractive in this case.

All variation in the dependent variable in Tables 1.2, 1.3, and 1.4 is linear,
but nonlinear effects of age, period, and cohort are not confounded with one
another in the same way as are the linear effects. Consider, for instance, the
nonlinear pattern of variation by age of the dependent variable in Table 1.6.
The only plausible explanation for the data is that, in the absence of compositional effects, they reflect nonlinear age effects, because the nonlinear
pattern of variation by age cuts across the different periods and cohorts. The
meaning of the data would also be clear if there were a nonlinear pattern of
variation by period that cut across the different ages and cohorts, or if there
were a nonlinear pattern of variation by cohort that cut across the different
ages and periods. If more than one kind of effect is reflected in the data, of
course estimation of the effects cannot be accomplished by simple inspection of cross-tabular data, but, as I illustrate in the next chapter, if all effects

are nonlinear, it may be possible statistically to estimate age, period, and cohort effects with reasonable accuracy. However, the fact that there is some deviation from linearity in the variation of the dependent variable by age, period, and cohort does not mean that the effects can be separated statistically. If any major component of the variation is linear, a statistical separation of the effects is impossible.

In view of the fact that the identification problem involves basic and widely understood mathematical principles and is involved in many kinds of research in addition to cohort analysis (such as estimating the effects of mobility or of husband-wife differences), it is surprising that the problem apparently was not recognized by a large majority of social scientists as recently as the 1960s, when Hubert M. Blalock, Jr., published a series of articles on it (e.g., Blalock, 1966, 1967). It was not until 1973 that the first widely cited article explicating the identification problem in cohort analysis was published (Mason, Mason, Winsborough, & Poole, 1973). The authors of that classic piece did a masterful job of explaining in simple language some of the logical and mathematical issues involved. Unfortunately, however, the article also raised unrealistic expectations about *solving* the identification problem through statistical analysis, and thus, it spawned a host of mainly ill-conceived attempts to separate the effects confounded in cohort data. Some of those are discussed in the next chapter.

2. STRATEGIES FOR ESTIMATING AGE, PERIOD, AND COHORT EFFECTS

The Mason, Mason, Winsborough, and Poole Method

The most commonly used method for estimating age, period, and cohort effects was introduced by Mason et al. (1973) in their classic article explicating the identification problem in cohort analysis. That method and its variants have continued to be widely used as competing methods have been introduced.

The Mason et al. method is seductive in its simplicity. Age, period, and cohort are each recoded into a set of dummy variables, with each dummy variable usually representing a range of 5 or 10 years. When the variables in a set are entered into a regression analysis as predictor variables, one variable in the set must be omitted to get the program to run. All that has to be done to get the program to run with age, period, and cohort dummy sets all entered is to omit one additional dummy variable, which can be from either the age, period, or cohort set. Omitting this additional variable introduces into the analysis the assumption that the effects of two variables omitted from the same set are equal—usually not a substantial distortion of reality, especially if the categories represented by the omitted dummy variables are adjacent.

This, of course, is not the only possible way to get a regression program to run with age, period, and cohort simultaneously entered as predictor variables. There are various ways in which the variables can be recoded or transformed to break the linear dependence among them and allow the program to run, that is, to make an identifying restriction. For instance, age and cohort can be entered as continuous variables while period is entered as a set of dummy variables. However, when this is done, the linear dependence is broken *in the statistical model only* and not in the real world, and thus, the obtained estimates of effects are not meaningful.

It follows that the simplifying assumption made when the Mason et al. method is used must not be selected haphazardly. Some researchers have apparently believed that because assuming equality of the effects of adjacent age, period, or cohort categories is rarely a gross distortion of reality, selection of almost any two adjacent categories will do—something that Mason et al. never claimed.

Many, if not most, researchers who have used the Mason et al. method or one of its variants have also overlooked a crucial part of the article in which it was introduced—a part unfortunately buried in a footnote:

> These pure effects [the hypothetical effects used to illustrate the method] have deliberately been made nonlinear in form. . . . We create our data in this way because perfectly linear pure effects are inherently ambiguous to interpret, and can be estimated equally well by the pure effect variable or by the two remaining variables in cohort analysis. (Mason et al., 1973, p. 248)[1]

In other words, the method is applicable only for estimating nonlinear effects.

How poorly the method performs when effects are linear is illustrated by the estimates in Table 2.1 of age, period, and cohort effects by four models in which different simplifying assumptions are used. To perform these analyses, I constructed a data file based on the assumption that there was a five-point per 10 years positive period effect on the dependent variable and a five-point negative cohort effect, and that the value of the dependent variable in 1950 among persons ages 20–29 was 50. The resulting data are those in Table 1.2. For this simulation experiment, I know what the effects are and can apply the Mason et al. method to the data to see how well it performs. I use four different models, with different simplifying assumptions (identifying restrictions) used in each one. The model estimates are in Table 2.1.

TABLE 2.1

Four Mason et al. Age-Period-Cohort Models Estimating the
Effects Reflected in Table 1.2 (Unstandardized Regression Coefficients)

Variable	Model	1	2	3	4
Constant		50.0	50.0	25.0	28.7
	Age				
	20-29	a	a	0.0	a
	30-39	5.0	5.0	0.0	a
	40-49	10.0	10.0	a	3.6
	50-59	15.0	15.0	a	6.1
	60-69	20.0	20.0	0.0	8.6
	70-79	25.0	25.0	0.0	11.4
	Year				
	1950	a	0.0	a	a
	1960	0.0	0.0	5.0	a
	1970	0.0	0.0	10.0	3.6
	1980	0.0	a	15.0	6.1
	1990	0.0	a	20.0	8.6
	2000	0.0	0.0	25.0	11.4
	Cohort				
	(Year of Birth)				
	1871-1880	0.0	0.0	50.0	25.0
	1881-1890	0.0	0.0	45.0	23.9
	1891-1900	0.0	0.0	40.0	21.5
	1901-1910	0.0	0.0	35.0	19.1
	1911-1920	0.0	0.0	30.0	16.8
	1921-1930	0.0	0.0	25.0	14.0
	1931-1940	0.0	0.0	20.0	11.8
	1941-1950	0.0	0.0	15.0	9.1
	1951-1960	a	0.0	10.0	6.5
	1961-1970	a	0.0	5.0	3.9
	1971-1980	0.0	a	a	a

a. Reference category; value set at zero.

The method gives grossly incorrect results unless the effects are assumed to be the same for two adjacent age levels, in which case it yields the correct estimates. If the equal effects assumption is applied to two years or two cohorts, the method yields a one-variable (pure age effects) solution, whereas assuming that the effects are equal for two age levels *and* two years results in a three-variable interpretation. The data themselves provide no clues as to which equality assumption or assumptions to use, because each assumption involves about the same degree of distortion of reality (although, of course, two assumptions introduce more distortion than does one). Obviously, correct estimation of effects with the method depends on a priori knowledge of what interpretation is correct. The method cannot tell the researcher whether a one-, two-, or one of the many possible three-variable solutions accurately represents reality. Nor can any other method, as I point out in Chapter 1.

However, because nonlinear age, period, and cohort effects are not confounded with one another in the same way as are linear effects, one might think that these effects usually can be estimated pretty well by the Mason et al. method. Another simulation experiment can provide relevant evidence. This time, I assume a pattern of nonlinear pure age effects, whereby the value of the dependent variable at ages 20–29 is 50, and each 10 years of aging from that level has effects of +2, +10, 0, –12, and –5, respectively, up to ages 70–79. This pattern of effects produces the data shown in Table 1.6. I again use four different Mason et al. models to estimate the effects reflected in the data (Table 2.2). For Model 1, the assumption is that the effects for ages 30–39 and 40–49 are the same, which is correct, and the model yields accurate estimates. For Model 2, the assumption is that the effects for two cohorts are the same, which is correct because there are no cohort effects, and all of the APC estimates are correct. For Model 3, the assumption is that the effects for two years are the same, and again, the correct simplifying assumption leads to correct estimates of all age, period, and cohort effects. Indeed, it can be shown that any correct simplifying assumption used in Mason et al. analyses of these data will yield accurate estimates of effects. For Model 4, the simplifying assumption of equal effects for ages 20–29 and 30–39 is not quite correct, although it is not grossly in error, and it leads to estimates that are vastly at odds with the real effects. Whereas a one-variable (pure age effects) solution is the correct one, the model shows a three-variable pattern of effects.

The lesson is clear: Even when the effects are nonlinear in the simple fashion shown in Table 1.6, the Mason et al. method provides correct estimates only when the simplifying assumption used is precisely correct. And it is rarely the case that a researcher can be certain that a simplifying assumption is not at least slightly in error.

When the effects are nonlinear in a more complex fashion, say, including nonlinear effects of two or all of the APC variables, the Mason et al. method or any statistical modeling is even less adequate. Even though nonlinear effects are not confounded with one another in the same way as are linear effects, they may be confounded with interactions among the APC variables. For instance, whereas nonlinear age effects are reflected in a statistical analysis as an interaction between period and cohort, not all period-cohort interactions are age effects. The simple pattern of nonlinear variation in

TABLE 2.2

Four Mason et al. Age-Period-Cohort Models Estimating the Effects
Reflected in Table 1.6 (Unstandardized Regression Coefficients)

Variable	Model	1	2	3	4
Constant		62.0	50.0	50.0	60.0
	Age				
	20-29	−12.0	a	a	a
	30-39	−10.0	2.0	2.0	a
	40-49	a	12.0	12.0	8.0
	50-59	a	12.0	12.0	6.0
	60-69	−12.0	0.0	0.0	−8.0
	70-79	−17.0	−5.0	−5.0	−15.0
	Year				
	1950	a	a	a	a
	1960	0.0	0.0	a	2.0
	1970	0.0	0.0	0.0	4.0
	1980	0.0	0.0	0.0	6.0
	1990	0.0	0.0	0.0	8.0
	2000	0.0	0.0	0.0	10.0
	Cohort (Year of Birth)				
	1871-1880	0.0	0.0	a	20.0
	1881-1890	0.0	0.0	0.0	18.0
	1891-1900	0.0	0.0	0.0	16.0
	1901-1910	0.0	0.0	0.0	14.0
	1911-1920	0.0	0.0	0.0	12.0
	1921-1930	0.0	0.0	0.0	10.0
	1931-1940	0.0	0.0	0.0	8.0
	1941-1950	0.0	0.0	0.0	6.0
	1951-1960	0.0	0.0	0.0	4.0
	1961-1970	0.0	a	0.0	2.0
	1971-1980	a	a	0.0	a

a. Reference category; value set at zero.

Table 1.6 makes it easy to judge that the variation reflects age effects, but more complex data defy interpretation by simple inspection.

Consideration of APC interactions leads to still another reason why the Mason et al. method is not as useful as many researchers have believed it to be. All major variants of the method are based on the assumption that age, period, and cohort effects are additive, that is, that age effects are the same for all periods and cohorts, period effects are the same for all ages and cohorts, and cohort effects are the same for all ages and periods. This assumption can be tested, and it rarely seems to be correct. When changes in attitudes and behavior occur in an adult human population, there is almost always greater change among the younger persons. The data themselves cannot distinguish between age-period and cohort-period interactions, but there are strong theoretical reasons to believe that as adults grow older, they become less responsive to the stimuli that bring about change in the population (see Glenn, 1974, 1980, and Alwin, Cohen, & Newcomb, 1991, for discussion of some likely reasons for this age effect). There usually is differential attitude change by age regardless of whether the change is in a liberal or a conservative direction—an issue I address in the next chapter. Furthermore, some age effects, including especially some associated with social roles rather than biological aging, have changed recently and thus vary among birth cohorts, an example being an increased expectation for middle-aged and elderly persons to be sexually active. Cohort effects may change with age as well, as exemplified by the dwindling of the economic and occupational advantages of a small birth cohort as it grows out of young adulthood and faces competition from younger cohorts. None of this complexity can be captured by additive APC models.

Just how inadequate additive APC models can be, as well as the necessity for using side information to interpret cohort data, is illustrated by examination of the data on percentage of women married in Table 1.1. Interactions abound in the data. Although the percentage married varied negatively with age at all dates, the regression of percentage married on age varies from −1.012 for 1968 to −.293 for 1998. Also, whereas the highest percentage married was at ages 25–34 in 1968, at the later dates, the highest percentage was at ages 35–44. Then, the regression of percentage married on year varies from −.696 for ages 25–34 to .299 for ages 65–74. The cohort that was ages 25–34 in 1968 experienced a decline in the percentage married as it aged to 35–44, but during the same stage of aging, the cohort that was ages 25–34 in 1978 had virtually no change in the percentage married, and the cohort that was ages 25–34 in 1988 experienced an increase.

Statistically modeling the effects underlying these interactions would be difficult even if the identification problem did not plague the effort. However, the data are not mysterious to family demographers familiar with changes in marriage, divorce, and longevity in the United States in recent decades. The relevant underlying trends are (a) a substantial increase in the typical age at first marriage from the late 1970s through the 1980s, (b) a steep increase in divorce from the mid-1960s through around 1980 that involved long-term marriages only to a minor extent, (c) a decrease in the death rates of middle-aged and older men that began in the 1980s and has continued, and (d) the maturing into the older age brackets of the cohorts with the highest lifetime rates of marriage. To anyone not familiar with these trends, no statistical manipulation of the data in Table 1.1 could lead to much insight into the meaning of the complex pattern of the data.

The Nakamura Bayesian Method

The Mason et al. method of cohort analysis is applied mindlessly and routinely only if the simplifying assumption (identifying restriction) is selected haphazardly, which is a misuse of the method. Other methods of cohort analysis, however, are designed to be used mechanically, that is, to be applied in exactly the same way regardless of what theory or side information predicts about age, period, and cohort effects.

Perhaps the most notable of these methods was developed by Japanese statistician T. Nakamura (1982, 1986) and introduced to American social scientists by Sasaki and Suzuki (1987). The invariant simplifying assumption with this method is that successive parameters change gradually. More specifically, the assumption is that ages, periods, and cohorts adjacent to one another are more similar in their effects than those more widely separated on the scale. Further elaboration is given by Sasaki and Suzuki (1989):

> The percentage difference for certain social phenomena between short survey time spans is smaller than the percentage difference for certain social phenomena between longer survey time spans. To illustrate, the percentage difference for certain social phenomena such as religious commitment between, say 20-year-old age groups and 30-year-old age groups is in general smaller than that between 20-year-old age groups and 40-year-old age groups. (p. 764)

Sasaki and Suzuki admit that the assumption that successive parameters change gradually is not always correct, but they claim it usually is. And they clearly believe that the method will always give correct estimates if the assumption is correct.

If Sasaki and Suzuki's view of their method were valid, the method would indeed be useful, because the assumption of gradual change can be tested empirically. One need only look at the relationship of any two of the APC variables to the dependent variable to see if it is monotonic. If it is, or if there are only minor reversals, the assumption would seem to be correct, or nearly so.

It can be shown, however, that the correctness of the assumption does not ensure the correctness of the estimates of age, period, and cohort effects. Sasaki and Suzuki (1989) applied their method to a set of cohort data very similar to those in Table 1.4, in which all variation by age and period was not only monotonic but perfectly linear. They concluded that the data reflected only cohort effects, although an infinite number of other interpretations of the data are logically possible, and some of those would be plausible for certain dependent variables. It appears that the method always chooses the simplest of the possible interpretations, at least as simplicity is operationally defined by the method. That interpretation may, more often than not, be correct, but "more often than not" is not good enough to support a recommendation that researchers apply the method and accept the estimates as correct. Nor should the Nakamura estimates be considered more likely to be correct than conclusions about the effects based on more informal, less rigorous methods.

Clearly, the estimates provided by the Nakamura method should always be viewed with skepticism, but the method may be useful if the estimates are evaluated with the use of theory and side information and through informal assessments of the patterns in the data. Consider, for instance, Dutch data on religious nonaffiliation reported and analyzed by Sasaki and Suzuki (1987), which are shown in Table 2.3. In a separate table, the authors show that in the total Dutch adult population, religious nonaffiliation increased monotonically from 1.8% in 1899 to 23.0% in 1969. Obviously, during the time covered by the data in Table 2.3, there were strong period influences for religious nonaffiliation. When such period influences are persistent, they almost always create an intercohort trend in the same direction as change in the total population. Hence, any seasoned cohort analyst who examines Table 2.3 will recognize the pattern of the data as reflecting largely positive period and cohort effects. Although nonaffiliation increased within each cohort that is represented for more than one time period in the table, there is no theory or evidence from any other source that predicts a decrease in religiosity or religious affiliation as a consequence of growing

TABLE 2.3
Percentage (Estimated) of Persons in the
Netherlands Without a Religious Affiliation, by Age and Year

Year	Age					
	20-29	30-39	40-49	50-59	60-69	70-79
1899	2.4	2.2	1.7	1.2	0.9	0.6
1909	5.8	5.6	4.2	3.2	2.4	1.6
1919	8.4	8.9	7.1	5.3	3.9	2.7
1929	15.0	16.2	14.1	11.0	8.1	5.8
1939	16.9	18.3	17.0	14.6	11.2	8.0
1949	18.0	19.5	19.0	17.5	14.3	10.4
1959	18.3	20.2	19.8	19.2	17.8	14.2
1969	24.7	23.0	23.3	23.1	21.4	19.2

SOURCE: Adapted with permission from Sasaki and Suzuki (1987), Table 4.

TABLE 2.4

Nakamura Bayesian and Ordinary Least Square Regression Estimates
of Age, Period, and Cohort Effects on Dutch Religious Nonaffiliation

	Nakamura Bayesian	Regression
Age		
20–29	−.0715	a
30–39	.0641	a
40–49	.0615	a
50–59	.0473	a
60–69	−.0063	a
70–79	−.0950	a
Period (Year)		
1899	−1.2301	a
1909	−.5361	1.352
1919	−.2775	2.269
1929	.2625	6.385
1939	.3305	7.406
1949	.3985	8.042
1959	.4355	8.732
1969	.6177	11.835
Cohort (Year of Birth)		
1820–1829	−1.2967	−5.686
1830–1839	−1.0314	−7.712
1840–1849	−.7699	−5.393
1850–1859	−.4908	−5.137
1860–1869	−.2230	−4.208
1870–1879	.0447	−2.578
1880–1889	.2913	a
1890–1899	.4545	2.286
1900–1909	.5195	3.814
1910–1919	.5575	4.536
1920–1929	.5754	4.678
1930–1939	.5539	4.081
1940–1949	.7742	6.579

SOURCE: Adapted with permission from Glenn (1989), Table 3.

a. Omitted from the regression.

older. Therefore, these intracohort shifts almost certainly are largely, if not entirely, period effects.

The Nakamura estimates of the effects reflected in these data are shown in Table 2.4, and they show precisely the pattern of effects that theory, side information, and informal assessment of the data indicate they should. In this case, the method apparently performs quite well. On the other hand, one can obtain similar estimates simply by assuming no age effects, converting period and cohort into dummy variables, and using ordinary least squares regression to estimate period and cohort effects on religious non-affiliation (see the last column in Table 2.4). It is not apparent in this case that the more complicated Nakamura method is superior to a simple regression analysis, although in other cases, it might be.

The Nakamura Bayesian method has not been used extensively, at least in the United States. Sasaki and Suzuki's claim that it provides an all-purpose solution to the identification problem in cohort analysis has apparently been generally rejected, and appropriately so. However, the method sometimes provides credible estimates of effects, and it probably comes about as close to being a good, all-purpose method of cohort analysis as is possible. It is curious, therefore, that although the Nakamura method has not caught on, the quest for a generally applicable statistical method of cohort analysis has continued.

The Quest Continues

As the manuscript for this monograph was nearing completion, at least a half-dozen manuscripts were being circulated in which a new method of statistically modeling APC effects is explicated and illustrated. The manuscripts were at the "do not cite" stage, and thus I cannot summarize their contents or reveal the identity of their authors. Suffice it to say that at least some of the authors were touting this new method as an all-purpose solution to the APC conundrum that will yield meaningful estimates when applied to any comparable repeated cross-sectional data gathered numerous times at short intervals. At least as the method is illustrated in the circulated papers, there is no provision for altering the analysis to take into account theory and side information , and there is no way to take into account the inherent ambiguity of a linear pattern of variation of the values of a dependent variable across the APC variables. In other words, at least some proponents of the method are claiming that it can accomplish the logically impossible.

The method may prove to be useful, however, if it yields approximately correct estimates "more often than not," if researchers carefully assess the credibility of the estimates by using theory and side information, and if they keep their conclusions about effects tentative.

Papers explicating and illustrating the method are likely to be published soon, perhaps before this monograph is published. My advice to researchers tempted to use the method is to wait until it is thoroughly vetted with simulation experiments that demonstrate how it performs in different circumstances. Or, researchers may do the simulations themselves.

When the simulations are done, it is important to keep in mind that a few successful cases of estimating the effects will not prove that the method is universally applicable. It is also important that complex, though plausible, combinations of effects be used for the experiments.

Age-Period-Cohort-Characteristic Models

A fairly recent innovation in the study of some aspects of cohort effects is the introduction of age-period-cohort-characteristic (APCC) models (O'Brien, 1989, 2000; O'Brien, Stockard, & Isaacson, 1999). These models do not include cohort but do include one or more "cohort characteristics," such as cohort size or some measure of family structure that varies by cohort. Although it is impossible to control age and period and let cohort vary, it is possible to control age and period and let the cohort characteristics vary, provided that the correlation of the characteristics with cohort is not too high. If the interest is only in estimating the effects of the selected cohort characteristics, and not of estimating the effects of age and period, and if the cohort characteristics do not bear a strong linear relationship to cohort, these models should serve their purpose reasonably well. It is extremely important to keep in mind, however, that these models are not true APC models (although they are sometimes represented as such) and in no sense provide a solution to the age-period-cohort conundrum.

It is unlikely that the cohort effects on any dependent variable result only from the cohort characteristics included in the models, and any remaining cohort effects are confounded in the model estimates with age and period effects. It follows that the estimates of age and period effects provided by the models are not meaningful. Furthermore, to the extent that the cohort characteristics are linearly related to cohort, their effects are also confounded in the estimates with age and period effects. Fortunately, in recent American data, the relationship of cohort size to date of birth (birth cohort) is largely nonlinear, and thus, APCC estimates of the effects of cohort size should be fairly accurate. However, some other cohort characteristics, such as the proportion of babies born out of wedlock and the proportion of cohort members who experienced a parental divorce before they reached adulthood, bear a monotonic, though not perfectly linear, relationship to cohort. Such characteristics have been used in APCC models, but the strong linear component of their relationship to cohort means that the estimates of their effects are not likely to be accurate.[2]

It is obvious that one can also conceive of age-characteristic-period-cohort models and age-period-characteristic-cohort models. Although I have never seen these terms used, there are instances of research in which period, cohort, and some characteristic loosely associated with age have been used simultaneously as independent variables, and there are also instances of the use of what could be labeled age-period-characteristic-cohort models.

Such models have the same strengths and limitations as APCC models, and with all such models, it is important to avoid independent variables that bear a strong linear relationship to the omitted APC variable. It is also important to remember that the estimates of the effects of the included APC variables are not meaningful.

Informal Means of Assessing APC Effects

The fact that statistical APC models cannot be relied on to provide accurate estimates of age, period, and cohort effects does not mean that one must give up on trying to distinguish the effects. Once the quest to separate the effects with precision and absolute certainty is abandoned, reasonable judgments about the effects can usually be made by using theory, side information, common sense, and various kinds of statistical analysis.

However, there is no formula, no cookbook approach, for distinguishing the effects that will work well in all cases. On rare occasions, only an eye-balling of crosstabular data is needed, as would be the case if one found variation in a dependent variable similar to that in Table 1.6. Similarly, there is only one plausible interpretation when there is a nonlinear pattern of variation in the dependent variable among cohorts that is consistent across age categories and periods, or if there is a nonlinear pattern of variation among periods that is consistent across age categories and cohorts. Usually, making reasonable judgments about the effects is more difficult and requires ingenuity, resourcefulness, insight, and good judgment on the part of the researcher.

Often, it is necessary to look at the data in a variety of ways and to bring in various kinds of side information in order to make fairly confident conclusions about age, period, and cohort effects (for examples, see Abramson & Englehart, 1995; Alwin, 1991; Converse, 1976; Glenn, 1994, 1998). Various kinds of statistical analysis, ranging from simple to complex, may be useful.

An example of a useful simple analysis is the statistical control of a variable that is known to intervene between either age, period, or cohort and the dependent variable. To illustrate, in the United States, the percentage of persons who grew up in rural areas varies considerably by birth cohort; thus, any variation by age in the dependent variable that is removed by controlling size of community of origin is probably a cohort rather than an age effect. Such preadult background variables as number of siblings and whether or not parents divorced also intervene between birth cohort and some dependent variables. Amount of education is another cohort-related variable, but the results of controlling it must be interpreted with caution in view of the fact that it apparently has had different effects in different birth cohorts (Alwin, 1991; Glenn, 1994). Such variables as marital status, parental status, employment status, and physical health may intervene between age and various dependent variables; level of prosperity, unemployment rate, and general political climate are examples of period-related variables that may be usefully controlled in cohort analyses.

Given the fact that informal cohort studies must be adapted to the particular problem at hand, I cannot describe and give illustrations of all of the useful techniques that have been applied. However, I report an analysis in which reported personal happiness is the dependent variable in order to illustrate some of the useful strategies.

An Illustration: A Cohort Analysis of Personal Happiness

The purpose of this study is tentatively to answer the following two questions:

1. In American society, what have been the typical effects on personal happiness of aging during adulthood?

2. How have these typical effects differed between men and women?

The source of data for this study is the American General Social Surveys (GSS) conducted from 1972 through 2002 by the National Opinion Research Center at the University of Chicago (Davis, Smith, & Marsden, 2002). The surveys were conducted annually from 1972 through 1978, in 1980, annually from 1982 through 1991, and in 1993, 1994, 1996, 1998, 2000, and 2002. On each survey, the respondents were asked the following question:

Taken all together, how would you say things are these days—would you say that you are very happy, pretty happy, or not too happy?

This 3-point ordinal scale has usually been treated in research as though it were an interval scale, which I do here for regression analyses. However, for aggregate-level analyses, I use two other measures, namely, (a) what I call the Happiness Index, which is the percentage of "very happy" responses minus the percentage of "not too happy" responses; and (b) a simple dichotomy of "very happy" versus other responses.

When the data from the 1972 through 2002 surveys are pooled, the relationship of the Happiness Index (HI) to age, for males and females separately, is that shown in Figure 2.1.

24

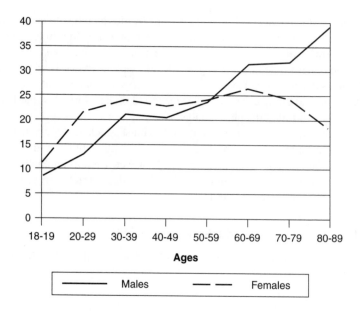

Figure 2.1   Happiness Index[a] by Age and Sex, United States

SOURCE: Pooled data from the 1972–2002 United States General Social Surveys (Davis et al., 2002).

a. Percentage of respondents who said they were "very happy" minus the percentage who said they were "not too happy."

For males, the relationship is monotonically positive, but for females, the HI is highest in late middle age, lower at the more advanced ages, and lowest in young adulthood. Up to the 50s, females reported greater happiness than males, but this difference is sharply reversed among the elderly.

As the reader should know at this point, these relationships of reported happiness to age may confound age, cohort, and attrition (compositional) effects. The most likely first interpretation of the data is that they show how the happiness of men and women changes as they grow older, but there are alternative explanations. For instance, the positive monotonic relationship of the HI with age in the case of men could have resulted from each successive cohort that reached adulthood being less happy than the one before it. More plausibly, the relationship of the male-female difference to age could have resulted from the gains women in American society have made in the past few decades, whereby younger cohorts of women are less disadvantaged relative to men than older ones.

TABLE 2.5

Happiness Index[a] by Age, Period,

and Sex, United States (*n* in parentheses)[b]

Age	1972-1982		1983-1992		1993-2002	
	Males	*Females*	*Males*	*Females*	*Males*	*Females*
18-19	4.4	9.0	11.9	12.6	11.2	13.4
	(250)	(211)	(168)	(151)	(143)	(134)
20-29	8.8	21.7	17.7	23.4	13.6	19.5
	(1,059)	(1,195)	(1,447)	(1,679)	(1,261)	(1,364)
30-39	19.3	26.9	21.9	22.0	22.2	23.5
	(1,218)	(1,495)	(1,343)	(1,680)	(1,416)	(1,666)
40-49	19.6	26.2	22.1	23.6	19.8	19.5
	(1,158)	(1,366)	(1,074)	(1,323)	(1,348)	(1,617)
50-59	23.4	25.8	23.3	20.1	25.2	25.8
	(1,183)	(1,319)	(803)	(935)	(917)	(1.078)
60-69	28.6	25.6	32.7	27.9	34.4	25.8
	(829)	(906)	(707)	(922)	(591)	(704)
70-79	30.5	25.6	30.7	26.1	34.7	21.6
	(471)	(510)	(407)	(583)	(427)	(555)
80-89	33.1	18.3	29.7	24.2	30.1	11.9
	(118)	(147)	(128)	(223)	(146)	(252)
Total	19.1	24.5	22.7	23.3	22.3	21.7
	(6,958)	(7,869)	(6,069)	(7,496)	(6,249)	(7,370)

SOURCE: Computed from the data file for the 1972-2002 United States General Social Surveys (Davis et al., 2002).

a. Percentage of respondents who said they were "very happy" minus the percentage who said they were "not too happy."

b. Number of respondents is approximate due to weighting. Data are weighted by number of adults in household divided by mean number of adults in GSS households.

When pooled cross-sectional data cover several years or decades, it is always useful to see if the indicated relationship was stable over the period of time covered. If it was not, the indicated changes can suggest how the data should be interpreted. Therefore, I tabulated the data in Table 2.5. Visual inspection of the data reveals no substantial changes in the relationship between age and reported happiness for either males and females—an impression that is confirmed by regressing the 3-point happiness scale on age for each sex and each period. The unstandardized regression for males is .004, .003, and 004 for the three time periods, and each coefficient is statistically significant at the .001 level on a two-tailed test. The comparable coefficients for females are .001, .001, and .000—none of which is statistically

TABLE 2.6
Percentage Very Happy,[a] by Period
and Sex, United States (n in parentheses)[b]

	Males	Females	Male-Female Difference
Period			
All Ages			
1972-1982	33.2	36.6	−4.4***
	(6,973)	(7,905)	
1983-1992	32.8	33.6	−0.8
	(6,084)	(7,523)	
1993-2002	32.4	32.7	−0.3
	(6,256)	(7,388)	
Change			4.1***
Ages 18-34			
1972-1982	24.7	34.8	−10.1***
	(2,641)	(2,935)	
1983-1992	28.3	31.6	−3.3*
	(2,286)	(2.712)	
1993-2002	27.6	31.4	−3.8**
	(2,113)	(2,328)	
Change			6.3***

SOURCE: Computed from the data file for the 1972-2002 United States General Social Surveys (Davis et al., 2002).

a. The other response alternatives are *pretty happy* and *not too happy*.

b. Number of respondents is approximate due to weighting. Data are weighted by number of adults in the household divided by mean number of adults in GSS households.

*$p < .05$ (two-tailed); **$p < .01$ (two-tailed); ***$p < .001$ (two-tailed).

significant. Therefore, it seems justified to interpret the pooled data without taking into account changes from 1972 to 2002 in the relationship between age and reported happiness.

Other changes from 1972 to 2002 are important, however, and must be taken into account. For instance, if the cohort explanation for the positive relationship of happiness to age among men were correct, the overall level of happiness of men should have declined from 1972 to 2002, with the decline being especially steep at the youngest adult age levels. The data in Tables 2.6 and 2.7 cast strong doubt on this interpretation, although they do not conclusively refute it. Those data show an increase in the mean reported happiness of men, especially at the 18–34 age level. Similarly, if the female advantage in happiness at the lower ages and the male advantage at the

TABLE 2.7
Regression (Unstandardized) of Happiness[a]
on Year (1972-2002), by Sex, United States

	Males	Females	Male-Female Difference
All Ages	.002***	−.001	.003***
Ages 18-34	.003***	−.001	.004**

SOURCE: Computed from the data file for the 1972-2002 United States General Social Surveys (Davis et al., 2002).

a. Happiness is measured on a 3-point ordinal scale treated as though it were interval.

$**p < .01$ (two-tailed); $***p < .001$ (two-tailed).

older ages were entirely or largely a cohort difference, females should have gained on males from 1972 to 2002, again especially among young adults. However, the data in Figure 2.2 and Tables 2.6 and 2.7 show the opposite trend; men gained on women, with the change apparently being greater among young adults than in the adult population as a whole. These interpretations depend on the assumption that there were no strong period effects that offset cohort effects—a reasonable assumption because period and cohort effects ordinarily (but not always) reinforce rather than offset one another in their contributions to trends in the total population.

This evidence that strongly suggests that the relationships of happiness to age shown in Figure 2.1 reflect primarily age effects can be supplemented usefully by data on the trends in reported happiness in birth cohorts as they grew older from 1972 to 2002. The usual procedure in tracing intracohort trends has been to take a fairly broadly defined birth cohort (e.g., a 5- or 10-year cohort) and trace it from the first to the last year for which there are data. A disadvantage of this procedure is that the older and younger members of the cohorts are not traced through the same ages. For instance, if one takes the cohort that was ages 20–29 in 1972 and traces it to 2002, the oldest members are tracked from age 29 to age 59, whereas the youngest members are followed from age 20 to age 50. This procedure is adequate for some purposes, but greater precision in discerning what happened at specific ages is achieved by tracing all members of each 5- or 10-year cohort through precisely the same ages. Of course, when this is done, the different cohort members are tracked through different periods of time, and if the data are confined to time points for which there are data for all cohort members (which is preferable in order to keep sample sizes large enough for reliable estimates), the span of time that can be covered is less. In the case of the 1972–2002 GSS data, the latter procedure allows for tracking each cohort for 20 rather than 30 years.

28

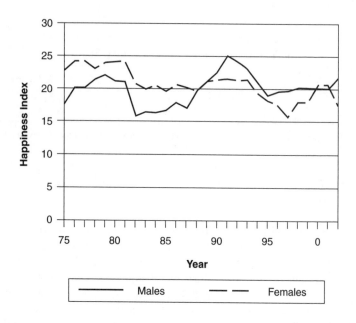

Figure 2.2   Happiness Index[a] by Period and Sex, 4-Year Running Means,[b]
United States

SOURCE: Computed from the data file for the 1972–2002 United States General Social Surveys (Davis et al., 2002).

a. Percentage of respondents who said they were "very happy" minus the percentage who said they were "not too happy."

b. Indicated years are the last years in the 4-year series.

In Figures 2.3 through 2.7, I show the intracohort trends in the Happiness Index for five 10-year birth cohorts, ranging from the 1915–1924 to the 1955–1964 cohort, which, in combination, are tracked from age 18 to age 77. If the age effect interpretation of the data in Figure 2.1 is correct, in each cohort, males should have experienced an increase in reported happiness, and the male trend should have been more favorable than the female one. The data in Figures 2.3 through 2.7 and in Table 2.8 generally conform to these predictions. Indeed, the indicated changes are greater than those needed to produce the relationships of age to reported happiness shown in Figure 2.1. Whereas the cross-sectional regression (unstandardized) of reported happiness (3-point scale) on age is .002 for males (Table 2.7), the corresponding per-year change shown by the intracohort trend data is .006 (mean for the five cohorts—computed from data in Table 2.8). Furthermore, the male-female difference in the regression coefficients for the

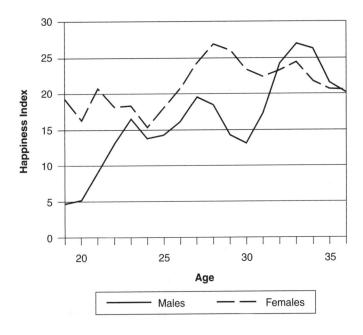

Figure 2.3   Happiness Index[a] for Persons in the 1955-1964 Birth Cohort as They Aged from 18 to 37, 3-Year Running Means[b]

SOURCE: Computed from the data file for the 1972-2002 United States General Social Surveys (Davis et al., 2002).

a. Percentage of respondents who said they were "very happy" minus the percentage who said they were "not too happy."

b. Indicated ages are the middle ages in the 3-year series.

cross-sectional data is .003 (Table 2.7), but the indicated mean per-year within cohort change in the male-female difference in happiness is .08 (computed from data in Table 2.8).

The most reasonable interpretation of these data is that the intracohort trends resulted from a combination of age and period effects—the period effects apparently resulting from the period influences suggested by the trends among young adults shown in Tables 2.6 and 2.7. In other words, it seems that recent period influences have increased the happiness of men relative to that of women, leading to changes within aging cohorts that add to same-signed age effects.

The trends among young adults shown in Tables 2.6 and 2.7 suggest period influences, but they also reflect cohort effects. The extent to which

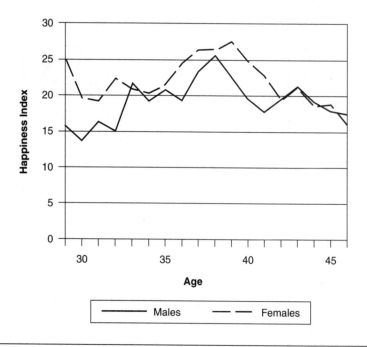

Figure 2.4   Happiness Index[a] for Persons in the 1945-1954 Birth Cohort as They Aged from 28 to 47, 3-Year Running Means[b]

SOURCE: Computed from the data file for the 1972-2002 United States General Social Surveys (Davis et al., 2002).

a. Percentage of respondents who said they were "very happy" minus the percentage who said they were "not too happy."

b. Indicated ages are the middle ages in the 3-year series.

these trends should be conceptualized as period effects or cohort effects is not clear, but acknowledging that they are at least partially cohort effects is important for interpreting the cross-sectional data on age and happiness. For instance, any intercohort trend toward greater happiness will reduce cross-sectional age differences created by positive age effects. This means that the positive age effects on happiness among men have almost certainly been somewhat greater than the cross-sectional data in Figure 2.1 indicate.

The bottom line is that in recent decades in the United States, the happiness of men apparently has increased monotonically as they have grown older, and because they have grown older, as suggested by the data in Figure 2.1, but positive age effects have very likely been greater than those suggested by the data. The happiness of women has apparently followed a nonmonotonic course as, and because, they have grown older, with the age

31

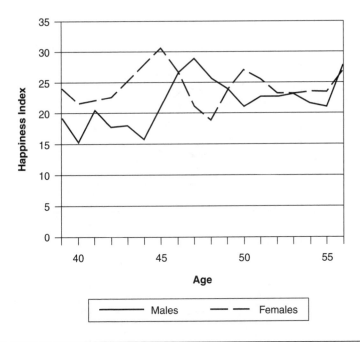

Figure 2.5   Happiness Index[a] for Persons in the 1935-1944 Birth Cohort as They Aged from 38 to 57, 3-Year Running Means[b]

SOURCE: Computed from the data file for the 1972-2002 United States General Social Surveys (Davis et al., 2002).

a. Percentage of respondents who said they were "very happy" minus the percentage who said they were "not too happy."

b. Indicated ages are the middle ages in the 3-year series.

effects being less favorable for women than for men at all, or almost all, age levels, and being negative at the older ages.

When elderly age ranges are included in a cohort study, it is always important to consider possible compositional effects due to differential mortality on the dependent variable. In this study, for instance, there is a possibility that a tendency for unhappy persons to die earlier than happy ones contributes to the apparent steep increase in the happiness of men from the 70s to the 80s. There may also be compositional effects for men at earlier ages and compositional effects for women that prevent their apparent decline in happiness in old age from being even steeper than it is.

The kind of evidence that would allow a definitive assessment of these possibilities is not available. The reported happiness of the elderly is

32

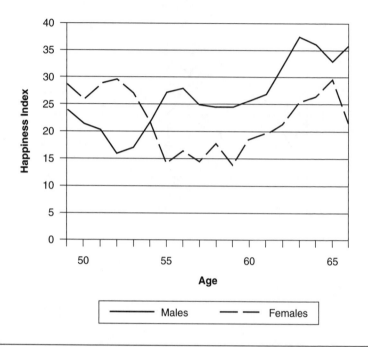

Figure 2.6   Happiness Index[a] for Persons in the 1925-1934 Birth Cohort as They Aged from 48 to 67, 3-Year Running Means[b]

SOURCE: Computed from the data file for the 1972-2002 United States General Social Surveys (Davis et al., 2002).

a. Percentage of respondents who said they were "very happy" minus the percentage who said they were "not too happy."

b. Indicated ages are the middle ages in the 3-year series.

somewhat predictive of their longevity (e.g., Palmore & Jeffers, 1971), but perhaps only because persons typically suffer a decline in health before they die, and because poor subjective health is related to low reported happiness. If so, the positive effects of mortality on the aggregate level of reported happiness should be largely or entirely offset by negative effects of predeath declines in health. What is needed but is not available to my knowledge is information on how lifetime levels of happiness are related to longevity. (The fact that subjective health declines with age makes the apparent positive age effects on happiness for men even more remarkable than they would be otherwise.)

It seems reasonable to conclude that differential mortality generally is not the source of the age and cohort patterns shown by the findings of this

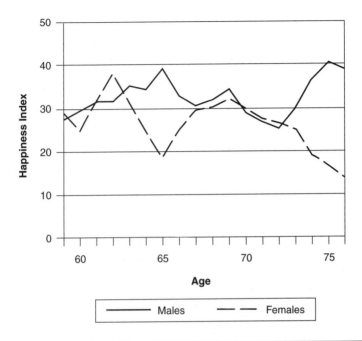

Figure 2.7   Happiness Index[a] for Persons in the 1915-1924 Birth Cohort as They Aged from 58 to 77, 3-Year Running Means[b]

SOURCE: Computed from the data file for the 1972-2002 United States General Social Surveys (Davis et al., 2002).

a. Percentage of respondents who said they were "very happy" minus the percentage who said they were "not too happy."

b. Indicated ages are the middle ages in the 3-year series.

study. However, it is an attractive explanation for the apparent sharp rise in happiness among men from the 70s to the 80s, if only because it is hard to think of any other reason for that finding.

The informal examination of the data I employ here provides convincing evidence of the direction and general nature of age effects on happiness for American men and women, respectively, for the past few decades. It does not allow precise estimates of the magnitude of those effects, which are confounded with cohort effects in the cross-sectional data and with period effects in the intracohort trend data. One could use statistical modeling to arrive at precise estimates, but there almost certainly would be considerable error in the estimates, which would be credible only if they conformed to the general pattern arrived at by informal means. And it is doubtful that

34

TABLE 2.8

Regression (Unstandardized) of Happiness[a]
on Age Within Each of Five 10-Year Birth
Cohorts as They Aged 20 Years, by Sex, United States

Birth Cohort (Range of Aging)	Males	Females	Male-Female Difference
1915-1924 (58-77)	.004	−.006*	.010*
1925-1934 (48-67)	.009**	−.004	.013**
1935-1944 (38-57)	.005	.000	.005
1945-1954 (28-47)	.002	−.003	.005
1955-1964 (18-37)	.009***	.003	.006*

SOURCE: Computed from the data file for the 1972-2002 United States General Social Surveys (Davis et al., 2002).

a. Happiness is measured on a 3-point ordinal scale treated as though it were interval.

*$p < .05$ (two-tailed); **$p < .01$ (two-tailed); ***$p < .001$ (two-tailed).

precise estimates that are unlikely to be precisely correct are any more useful for most purposes than knowledge of the direction and general pattern of the effects.

In a full study of age and happiness, it would, of course, be important to try to give reasons for the estimated effects and to provide relevant evidence if possible. Here, I simply point out that one of the more likely reasons for the apparently differing pattern of age effects on happiness for men and women is the differing pattern by sex of marital status by age. At all ages and for both sexes, persons who are married report, on average, considerably higher happiness than persons who are not married. This association may result in part from the nature of selection into and out of marriage, but it almost certainly results to some extent from a positive effect of being married on happiness. At the younger adult ages, women are more likely to be married than men, but at the older ages, men are more likely to be married than women. According to the 1972–2002 United States General Social Survey Data, the male-female difference in the percentage of persons married bore an almost perfect linear relationship to age within the 19 through 76 age range ($r = .95$ when 1-year age levels are the units of analysis). The male-female difference in the percentage of persons who said they were "very happy" also bore a strong, although less nearly perfect, linear relationship to age ($r = .792$), and the male-female differences in percentage married and percentage "very happy" correlate with a value of .823. Again using 1-year age levels as units of analysis and restricting the analysis to the age range of 19 through 76 (to keep sample sizes within cells at 100 or above), the regression (unstandardized) of the male-female difference in percentage "very happy" on age is .253 ($p < .01$ on a two-tailed test) but is reduced to .033 (a reduction of 87%) by control of the male-female difference in percentage married. When age is controlled, the regression of the male-female difference in happiness on the difference in percentage married is .263 ($p < .01$ on a two-tailed test).

Another way to approach this issue is simply to regress reported happiness on age for males and females respectively and see how the difference between the male and female regression coefficients is affected by adding marital status (married vs. not married) to the regressions as a covariate. When the regressions are run with the entire 1972–2002 GSS data file, the unstandardized bivariate coefficients are .004 and .001, for males and females, respectively, for a difference of .003 ($p < .001$ on a two-tailed test). Adding marital status to the equation reduces the male coefficient to .001 and leaves the female coefficient at that level (because the *linear* relationship of age to marital status among women is virtually nil).

These data indicate, but do not conclusively prove, that male-female differences in the age pattern of marital status have largely caused the apparent male-female differences in age effects on happiness.

3. USE OF COHORT ANALYSIS
FOR UNDERSTANDING CHANGE

I point out below that cohort analysis can be usefully applied only with data from populations into which and out of which there is little movement except through birth (or some other form of creation), death, or aging. In such a relatively "closed" population, change in the characteristics of the population between Times A and B occurs in three major ways, namely, through (a) addition of new individuals to the population through birth or aging, (b) subtraction of individuals through death or aging, and (c) changes, except those that result from aging, in the characteristics of individuals in the population at both Times A and B. The first two sources together constitute *cohort succession* and the third is *intracohort change.*[3]

In the first edition of *Cohort Analysis* (Glenn, 1977), I suggested that it would often be useful to decompose total population change into that resulting from the different sources. I reported some crude methods to do a rough decomposition, and I called for the development of more refined methods to do that. At least one analyst, Glenn Firebaugh (1989, 1990,

1992, 1997), has answered the call and devised several related methods of decomposing change into its sources.

I do not describe Firebaugh's techniques here because I now believe that neither they nor any similar means of decomposition is very helpful for understanding change (for a view similar to mine, see Rodgers, 1990). The decomposition is meaningful only in the absence of age effects on the dependent variable, and even on the rare occasions when one can be highly confident that there are no age effects, the results of the decomposition are too dependent on the length of time covered to be very useful.

How age effects render decomposition of population change meaningless is illustrated by the following hypothetical case. Suppose that in a population of persons ages 25–64, each 10 years of aging had an effect of +10 on the dependent variable, the age distribution did not change between Times A and B, there was a period effect of +10 from Time A to Time B, and cohort succession had no effect during the time covered. Age effects do not contribute to population change if the age distribution stays constant, but they do contribute to intracohort change, adding to that caused by period influences. Thus, in this hypothetical case, intracohort change will be more than enough to account for total population change. If Firebaugh's or any similar methods are applied, they will show a substantial negative contribution of cohort succession to population change, even though, in fact, cohort succession had no effect.

An additional problem is that cohort succession can contribute little to overall population change in a short period of time, whereas if the time is long enough, any decomposition method will attribute all of the change to cohort succession. Consider, for example, the population of persons ages 25–64. After 40 years, none of the original members of the population will be left, and thus, no portion of the observed population change can be attributed to intracohort change. Varying the amount of time covered moderately, say, from 25 to 35 years, will often appreciably affect the portion of change allotted to cohort succession and may affect interpretation of the observed population change (although it should not do so).

Nevertheless, the cohort approach to understanding social, cultural, and political change is almost always useful if appropriate data are available. It can be used, for instance, to identify change that persists because of cohort succession for several decades after the influences responsible for the change have ended. If, as theory and considerable evidence suggests, period influences typically have greater effects among younger adults than among older ones, then strong period influences will tend to leave substantial intercohort differences in their wake.

This was apparently the case with the influences associated with the "sexual revolution" of the 1960s and early 1970s in the United States. In the

TABLE 3.1
Regression (Unstandardized) of Sexual
Restrictiveness-Permissiveness on Age,[a] by Period, United States

Period	Premarital	Extramarital
1972–1976	−.022***	−.010***
1977–1981	−.022***	−.007***
1982–1986	−.015***	−.004***
1987–1991	−.019***	−.004***
1992–1996	−.016***	.001
1997–2002	−.014***	.000

SOURCE: Computed from the data file for the 1972–2002 United States General Social Surveys (Davis et al., 2002).

a. Negative coefficients indicate that older persons have more restrictive attitudes than do younger ones.

***$p < .001$ (two-tailed).

mid-1970s, by which time the so-called revolution had largely run its course, there were rather large age differences in responses to questions asked on the General Social Surveys about the "wrongness" of different kinds of sexual behavior, with younger adults being substantially more permissive than older ones (see Table 3.1), especially with respect to premarital sex. These differences could have reflected age effects, whereby growing older tends to make persons more restrictive in their sexual attitudes, but for reasons pointed out below, it seems likely that they were largely the result of differential response by age to the stimuli of the sexual revolution.

These intercohort differences in the mid-1970s are useful in understanding an apparent paradox, namely, opposite-signed trends in premarital and extramarital sex attitudes from the mid-1970s through the end of the 20th century (Figure 3.1). Among all adults in the United States, attitudes toward premarital sex became more permissive and attitudes toward extramarital sex became more restrictive, both trends being statistically significant ($p < .001$ on a two-tailed test). These differing trends seem paradoxical because the two kinds of sexual attitudes are correlated with one another ($p = .301$ in the 1972–2002 General Social Survey data file) and relate to many other variables in a similar fashion. Therefore, it would be strange if period influences on these two kinds of attitudes were in opposite directions over the last quarter of the 20th century.

It is very unlikely that these period influences were in opposite directions. As I point out above, the trend in period influences is often reflected in trends in the characteristics of young adults, and among persons ages 18–29, attitudes toward both premarital and extramarital sex became more

38

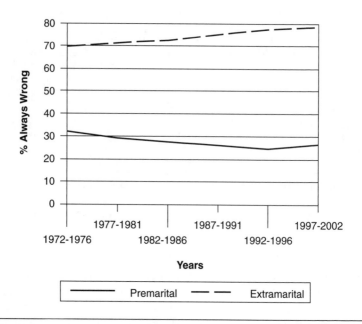

Figure 3.1   Percentage of Persons Age 18 and Older Who Said Premarital and
Extramarital Sex Are "Always Wrong," by Period, United States

SOURCE: Computed from the data file for the 1972-2002 United States General Social Surveys (Davis
et al., 2002).

restrictive from 1972 through 2002 (Figure 3.2). The change was much
greater for extramarital sex attitudes, but both trends are statistically significant ($p < .05$ for premarital sex and .001 for extramarital sex).

It is almost certain, therefore, that in the total adult population, attitudes
toward premarital sex continued to become more permissive only because
of cohort succession, the effects of which were strong enough to offset the
rather weak period influences toward restrictiveness. In contrast, period
influences toward restrictiveness in regard to extramarital sex were apparently much more than strong enough to offset the rather weak effects of
cohort succession. There may or may not have been age effects toward
restrictive attitudes, but if there were, those effects seem to have been over-
shadowed by the effects of cohort succession and period influences.

It is always useful, of course, to look at the trends within cohorts during
the time under consideration, and those trends within the three 10-year birth
cohorts that can be traced from 1972 to 2002 are shown in Table 3.2 and in
Figures 3.3 and 3.4. I use a dichotomous measure ("always wrong" vs. all
other responses) of sexual permissiveness-restrictiveness for the figures and

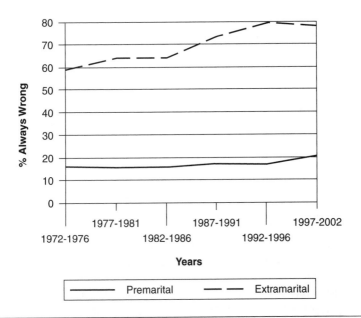

Figure 3.2   Percentage of Persons Ages 18-34 Who Said Premarital and
Extramarital Sex Are "Always Wrong," by Period, United States

SOURCE: Computed from the data file for the 1972-2002 United States General Social Surveys (Davis
et al., 2002).

for the logistic regression analysis reported in the bottom half of Table 3.2,
and I treat the 4-point ordinal scale as though it were the interval for the
ordinary least squares regression analysis reported in the top half of the
table.

The meaning of the premarital sex attitude intracohort trend data is not
entirely clear. Only the youngest cohort shows a statistically significant
trend, although both of the other cohorts apparently shifted toward restrictiveness between 1992–1996 and 1997–2002. I (Glenn, 2003), as well as
Harding and Jencks (2003), have interpreted similar data (not including the
most recent data used here) to suggest an age effect toward restrictiveness
from the earliest stage of adulthood to the 30s, but the similar changes in
all three cohorts toward the end of the period covered probably resulted
from period influences (see Figure 3.3). At any rate, the lack of trends
toward permissiveness within the cohorts, combined with the lack of such
a trend among young adults, indicates that the trend toward permissiveness
in the total adult population resulted entirely from cohort succession and
not at all from period influences.

TABLE 3.2

Regression (Unstandardized) of Sexual Restrictiveness-Permissiveness
on Year (1972-2002), by Birth Cohort, United States

Cohort (Year of Birth)	Premarital	Extramarital
Ordinary Least Squares[a]		
1925–1934[c]	–.002	–.008***
1935–1944[d]	.002	–.008***
1945–1954[e]	–.012***	–.011***
Logistic[b]		
1925–1934[c]	.004	.029***
1935–1944[d]	.006	.034***
1945–1954[e]	.030***	.035***

SOURCE: Computed from the data file for the 1972–2002 United States General Social Surveys (Davis et al., 2002).

a. Negative coefficients indicate a trend in a restrictive direction.

b. Responses are dichotomized *always wrong* versus all others. Positive coefficients indicate a trend in a restrictive direction.

c. This cohort aged from ages 38–47 to 64–73 during the time covered by the data.

d. This cohort aged from ages 28–37 to 54–63 during the time covered by the data.

e. This cohort aged from ages 18–27 to 44–53 during the time covered by the data.

***$p < .001$ (two-tailed).

The meaning of the intracohort trends in attitudes toward extramarital sex is clearer. There were strong and statistically significant trends toward restrictiveness in all three of the birth cohorts (Table 3.2 and Figure 3.4). In view of similar-size trends in the total adult population and among young adults (Figures 3.1 and 3.2), these changes were almost certainly largely in response to period influences, although some tendency for persons to become more restrictive in their extramarital sex attitudes as they grow older cannot be ruled out. The pro-restrictive period influences had to be strong enough not only to produce the observed trend in the total population, but also to offset any pro-permissive effects of cohort succession during the early part of the period covered by the data. Such pro-permissive effects are suggested by the positive relationship between age and restrictiveness shown for 1972–1976 through 1987–1991 in Table 3.1.

By the mid-1990s, however, the relationship between age and extramarital sex attitudes had disappeared, according to the data (Table 3.1). This change came about partly because younger persons shifted more in a restrictive direction than did older ones (Table 3.2), but it came about also because new cohorts that entered adulthood resembled older adults in their attitudes. Overall, these findings are consistent with the hypothesis that

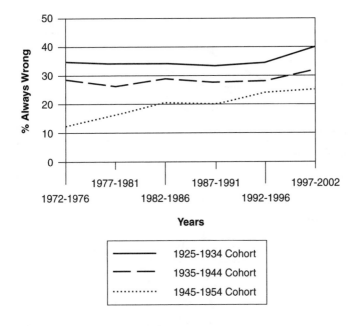

Figure 3.3 Percentage of Persons Who Said Premarital Sex Is "Always Wrong," by Birth Cohort and Period, United States

SOURCE: Computed from the data file for the 1972-2002 United States General Social Surveys (Davis et al., 2002).

younger adults are more responsive to stimuli for change than older ones, but the trend lines in Figure 3.4 do not show a monotonic convergence of the three birth cohorts. Furthermore, it is possible that the oldest cohort changed less than the younger ones only because of ceiling effects, that is, because persons in that cohort were at such a high level of restrictiveness that the measure used was not very sensitive to further movement in a restrictive direction.

When a dichotomous measure of a continuous variable is used, as when the responses to the sexual permissive-restrictive questions are collapsed into "always wrong" versus all other responses, it is useful to do a logit transformation of the percentages to "correct" the data for ceiling and floor effects. This transformation consists of converting percentages into odds and doing a logarithmic transformation of the odds. This procedures gives greater weight to percentage differences near zero or 100 than to those near 50 and is based on the assumption that the distribution of the continuous

42

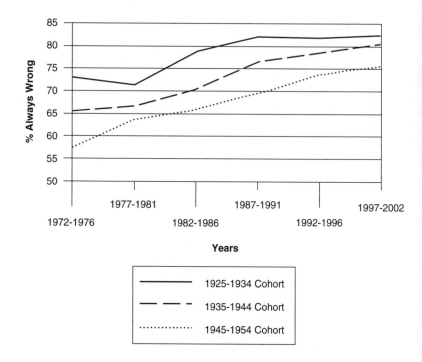

Figure 3.4   Percentage of Persons Who Said Extramarital Sex Is "Always Wrong," by Birth Cohort and Period, United States

SOURCE: Computed from the data file for the 1972-2002 United States General Social Surveys (Davis et al., 2002).

variable approximates normality. The accuracy of the "correction," of course, depends on the correctness of this assumption.

The logits that correspond to the percentages in Figure 3.4 are shown in Figure 3.5.

The data still show a net reduction in the intercohort differences from 1972–1976 to 1977–2002, but they also still show divergence among the cohorts from 1977–1981 to 1987–1991. Thus, as is often the case, the findings are "messy" in the sense of failing to give strong and consistent support to the hypothesis. There are several possible reasons, aside from sampling error, for the failure of the data to fall into the simple hypothesized pattern. For instance, age effects may be introducing complexity, or perhaps older persons respond almost as much to stimuli for change as younger ones but not as rapidly.

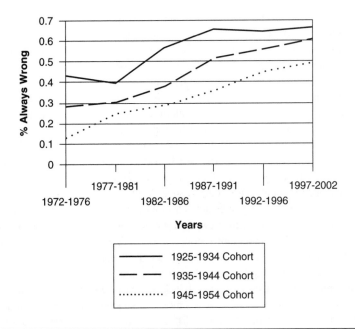

Figure 3.5   Logit Transformation of "Always Wrong" Responses to Extramarital Sex Question, by Birth Cohort and Period, United States

SOURCE: Computed from the data file for the 1972-2002 United States General Social Surveys (Davis et al., 2002).

4. DATA REQUIREMENTS AND AVAILABILITY

Data Requirements

Cohort analysis requires data collected at two or more points in time, and the most useful studies use data gathered at several times over a span of decades. Obviously, cohort studies rarely, if ever, follow the textbook sequence of planning the research, collecting the data, and doing the analysis. Rather, they are constrained by the existence and availability of appropriate data gathered by other researchers for other purposes. In other words, they are secondary analyses (Kiecolt & Nathan, 1985). Most of the usable data are from sample surveys, although some, such as those from national censuses, are from studies designed to cover all individuals (persons, households, or whatever) in a population.

A crucial requirement for cohort analysis is that the studied population be approximately "closed," that is, one into which and out of which there is little movement except through birth (or some other form of creation) and death, or through aging if an age-delineated subpopulation, such as ages 25–64, is used. Movement into and out of a population by other means can create a false appearance of change in aging cohorts, or it can mask change that has occurred. Therefore, the populations of regions, states (in the United States), communities, and similar geographic areas are not appropriate for cohort studies, nor are married persons, employed persons, or members of any population with highly permeable boundaries. Although no human population is completely closed, the adult populations of nation-states are usually reasonably suitable for cohort research, the exceptions being countries that have experienced very substantial immigration and/or emigration. Gender, racial, ethnic, and sometimes even religious subpopulations are also sufficiently closed. Of course, any subpopulation delineated on the basis of pre-adult background characteristics is closed during adulthood.

A second major requirement for cohort studies is that the data gathered at different times must be comparable. In the case of survey data, the question stems and response alternatives must be the same for all points in time unless there is evidence that any differences do not produce variation in response patterns. What seem to be minor differences in question wording can produce substantial differences in responses, as can differences in the order in which response alternatives are presented. Consider, for instance, a question about life satisfaction in which respondents are asked to rate themselves on a 7-point scale varying from *completely satisfied* to *completely dissatisfied.* The responses may vary according to whether the satisfied or dissatisfied alternative is presented first. They may also differ according to whether the intermediate points on the scale are identified verbally or just by numbers.

Even data from identically worded questions with identical response alternatives may not be comparable. Data collected by different survey organizations are likely to differ because of differences in sample design, interviewer training and supervision, and coding procedures (*house effects*), and data from different kinds of surveys (face-to-face, telephone, self-administered) are not strictly comparable (*mode effects*). More important, questions asked earlier, especially immediately before, can markedly affect how respondents answer survey questions (*question order* or *questionnaire context effects*). For instance, responses to a life satisfaction question apparently differ according to whether or not the question is preceded by questions asking about satisfaction with specific domains of life.

It is obviously important to use comparable data for tracing trends within cohorts or trends within age levels as succeeding cohorts occupy those levels. It is less obvious that even comparable data can give the appearance of a trend that has not occurred or mask a real trend. This fact is dramatically illustrated by the data in Figure 4.1, which show the inter-cohort trend in the life satisfaction of high school senior girls as indicated by the responses to two different life satisfaction questions used on different forms (used for different subsamples) of the Monitoring the Future Surveys conducted by the Institute for Social Research at the University of Michigan.

On Form 1 of the annual surveys, the students were asked to rate their satisfaction or dissatisfaction with various aspects of life on a 7-point scale varying from *completely satisfied,* which was listed first, to *completely dissatisfied.* The midpoint of the scale was identified as *neutral,* and other points on the scale were identified by numbers only rather than by numbers and verbal identifiers. After questions about 11 specific aspects of life, varying from "Your educational experiences" to "The way you spend your leisure time," the students were asked to rate "Your life as a whole these days." On Form 2 of the surveys, an independently drawn subsample of students was asked, after completely unrelated questions, "How satisfied are you with your life as a whole these days?" for which the response alternatives were seven levels of satisfaction and dissatisfaction varying from *completely dissatisfied* (listed first) to *completely satisfied,* with intermediate points on the 7-point scale identified by both numbers and verbal labels such as *quite satisfied* and *somewhat satisfied.* For Figure 4.1, I scored the responses to both questions with +3 for *completely satisfied* to –3 for *completely dissatisfied,* with 0 indicating the midpoint of the scale. The mean responses by year for each form are plotted in the figure.

Although the Form 1 and Form 2 questions are nominally equivalent, not only are the mean responses different, but the trends over the 25-year period are in opposite directions. When the surveys are used as the units of analysis, the correlation of the mean responses to the Form 1 question with year is –.458, which is statistically significant ($p < .05$ on a two-tailed test), whereas the correlation of the mean responses to Form 2 with year is +.793 ($p < .01$ on a two-tailed test). Obviously, both correlations cannot accurately indicate the real recent trend in the life satisfaction of female high school seniors in the United States.

The reason for the opposite-signed indicated trends is not apparent, because the two versions of the life satisfaction question differ in three different ways. First, and probably most important, the Form 1 life satisfaction question was preceded by questions about satisfaction with 11 specific domains of life. Second, the Form 1 question presented *completely satisfied* first and *completely unsatisfied* last among the response alternatives. And third, the Form 1 question used only numbers to identify four of the seven

46

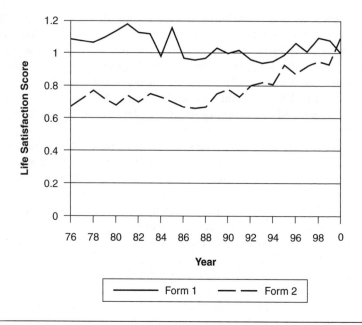

Figure 4.1 Life Satisfaction Scores[a] for High School Senior Females, by Year, United States

SOURCE: The 1976-2000 Monitoring the Future Surveys of High School Seniors (Bachman, Johnston, & O'Malley, 1976-2000).

a. Life satisfaction is measured on a 7-point scale ranging from *completely satisfied* (scored +3) to *completely dissatisfied* (scored –3).

points on the scale, whereas the Form 2 question used verbal labels for all seven points.

It is likely, though not certain, that the differing trends shown by the responses to the two questions resulted from question order effects on the Form 1 responses, because the responses to some, though not all, of the preceding domain-specific satisfaction questions on Form 1 also indicated a downward trend. That is, the preceding questions probably led respondents to think about some aspects of their lives on which satisfaction had declined. A trend in a variable treated by an earlier question may create the appearance of a trend, or mask a trend, in the variable of interest if there are question order effects on the latter.

The lesson to be learned from this illustration is that measures of dependent variables likely to be subject to substantial question order effects usually should be avoided in cohort studies. If such measures are used, the results of the analyses should be interpreted with great caution.

Even data gathered by the same research organization by the same mode (e.g., face-to-face interviewing) at different times may be incomparable because of changes in sample design. As I point out below, an abundance of American public opinion poll data from the late 1930s, 1940s, and 1950s might appear to be a valuable resource for trend and cohort studies, but unfortunately, the sampling used by the early opinion pollsters renders those data incomparable with most later data, including even the data gathered by the same organizations. The early pollsters all used some variant of quota control sampling, whereby quotas were set for major demographic categories based on age, sex, region, and so forth. Each interviewer was given a set of quotas; for instance, an interviewer might be asked to interview 10 males, 10 females, six adults under age 35, eight persons ages 35–59, and six persons age 60 or older. Within the limits set by the quotas, interviewers were given discretion in selecting respondents—a fatal flaw that led to substantial biases. By the late 1950s, most opinion pollsters were using more sophisticated sampling, although not the full probability sampling being used by the U.S. Census Bureau and such academic survey organizations as the Survey Research Center at the University of Michigan. Typically, a multistage area sampling procedure similar to that used for full probability sampling was used down to the block level, but quotas were used to select respondents in the final stage. This kind of sampling, variously called *modified probability sampling* or *probability sampling with quotas,* differed crucially from quota control sampling in that interviewers followed a set of rules to select respondents and had no discretion in choosing persons to be interviewed. Strictly speaking, data gathered by this method are not comparable with either data from quota control samples or full probability samples, but apparently, they are more nearly comparable with the latter.

The greatest change in sampling among the major American opinion pollsters occurred in the American Gallup Organization, which, in its early decades, was known as the American Institute of Public Opinion. Gallup first gained prominence by correctly predicting the outcome of the 1936 presidential election, and for the next couple of decades, the organization had a sharply political focus and a preoccupation with predicting the outcome of national elections. Therefore, the early Gallup samples were designed to represent the electorate rather than the total adult population, there being a marked underrepresentation of demographic categories with low voter turnouts, such as women, persons with less than a high school education, and Blacks. The Gallup samples were modified to represent the total adult population beginning in the early 1950s—a change that was complete by about 1958. This change in sample design alone could have produced changes of up to 8 or 10 points in the percentage of respondents selecting some response alternatives.

At a time when the Gallup data constituted a large proportion of the American survey data appropriate for trend and cohort studies, I suggested multiple standardization procedures to make the earlier and later Gallup data more nearly comparable (Glenn, 1975). Occasionally, these may still be worth applying, but the growth of an abundance of high-quality academic survey data gathered at different times on a large number of topics of interest to social scientists makes the Gallup data less attractive than they once were. Therefore, I forgo discussion here of techniques to deal with problems of comparability in the early and later American Gallup data. Most cohort analysts will use data in which these problems are not present.

Less serious than the problems of comparing early and later Gallup data are the problems created by all major polling organizations changing their usual mode of questionnaire administration from face-to-face to telephone interviewing, typically in the late 1970s or early 1980s. The problems posed by the shift to telephone interviewing are not serious enough to discourage use in the same study of early and later poll data from the same polling organization, but the researcher needs to know when the transition to telephone interviewing took place and should look for sudden changes in responses to repeated questions that occurred at that time.

A problem of comparability that has usually been ignored by cohort analysts (including me) arises from the fact that very young adults are not as adequately represented in most American national surveys as are older adults. Virtually all of the major surveys sample only the noninstitutionalized population and thus do not include persons who live in "group quarters." Group quarters include prisons, hospitals, military barracks, and college dormitories—all of which, except hospitals, are occupied mainly by young adults. The biases introduced by this underrepresentation of certain kinds of young adults are, to some extent, offsetting, because those excluded from the samples range from very underprivileged persons to upper-middle-class college students. However, for some variables, the net bias could be more than negligible.

Researchers who do cohort analyses should always consider this potential problem and exercise care in interpreting indicated changes in dependent variables between the earliest and slightly later stages of adulthood. However, for most studies, the problem does not appear to be serious. The extent of the incomparability of cohort samples from very young adulthood and those from older ages can be assessed by looking at indicated intracohort change in background characteristics that rarely change as persons grow older. For these variables, any indicated intracohort changes not the

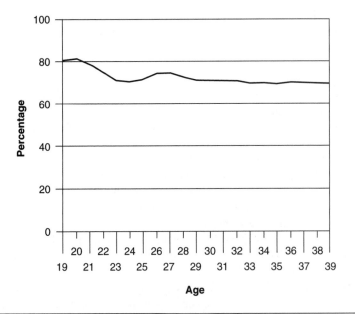

Figure 4.2   Percentage of Respondents' Mothers Who Had No More Than a High School Education, 1955-1964 Birth Cohort, by Age, United States, 3-Year Running Means[a]

SOURCE: Computed from the data file for the 1972-2002 United States General Social Surveys (Davis et al., 2002).

[a] Indicated ages are the middle ages in the 3-year series.

result of sampling error must be the result of sampling incomparability. For this assessment, it is useful to use the procedure illustrated earlier in Chapter 2 of reported happiness and to trace all 1-year cohorts through the same age range rather than through the same time period. I have used this procedure for several preadult background characteristics and have never found substantial indicated changes as the cohorts have grown older. An example is presented in Figure 4.2, in which the 1955–1964 birth cohort in the United States is traced from very young adulthood through the 30s in regard to having a mother who completed no more than a high school education. The indicated percentage with this characteristic dropped moderately from the late teens into the mid-20s, probably as a result of college students returning to the sampled population, and then it remained almost stable through the 30s. The small change shown could not have much effect on the results of a cohort analysis. However, researchers with no special interest in age effects among very young adults may use age 25 rather than age 18 as the youngest age included in their analyses in order to reduce the probability of bias from age-related sampling incomparability.

Data Availability

Most of the data upon which I draw for illustrations in this edition of *Cohort Analysis* are from the American General Social Surveys conducted either every year or every other year from 1972 through 2002. There are dozens of series of repeated questions on the GSS that I do not use, of which a few cover the full span of time from 1972 to 2002, many others start with 1973, and numerous others cover shorter spans of time. Some of the repeated questions were asked on every survey, and others appeared in a rotating pattern, such as being asked on two successive surveys, being left off of the next survey, and then returning for two more successive surveys. In recent years, the GSS has been conducted every other year, but with a sample twice as large as that used earlier and with many questions being asked of only a portion of the respondents.

Some of the longer series of questions on the American GSS include those about amount of satisfaction derived from various aspects of life, confidence in major institutions, ideological orientation, political party identification, voting in presidential elections, job satisfaction, work attitudes, alienation, voluntary association memberships, desirable child qualities, attitudes toward abortion, vocabulary, religious preference, religious beliefs, racial attitudes, smoking, and drinking.

The American General Social Surveys, in common with most other face-to-face surveys and many telephone surveys, sample households rather than individuals, and therefore, the data must be weighted by number of adults in the household in order to make them representative of the noninstitutionalized adult population. (The unweighted data overrepresent adults in one-adult households and underrepresent those in multiple-adult households.) A fractional weight rather than the raw number of adults in the household is typically used in order to keep the number of respondents shown by the analysis output close to the real number of respondents. That weight is constructed by dividing the number of adults in the household by the mean number of adults in GSS households (which is 1.94 for the 1972–2002 file). The weight can be computed for each survey but is usually computed for the file containing data from all of the surveys. A weight variable is included in the data set for oversamples of blacks on the 1982 and 1986 surveys, but as the GSS file gets bigger and the black oversamples become a very small proportion of the total, weighting for the oversamples now usually makes little difference. Statistical purists can weight the data to give equal weight to each survey, but this is a nicety that is rarely applied and usually makes a negligible difference in analysis results.

The GSS data are available from multiple sources, including on CD-ROM from the Roper Center for Public Opinion Research and online at the Web site for the Inter-university Consortium for Political and Social Research. Hard-copy codebooks can be purchased from the Roper Center.

The American National Election Studies (ANES) provide even longer series of repeated questions than do the General Social Surveys, with a few series, such as that on party identification, going back to the first ANES in 1948. Several other series start with 1952, and many cover a longer span of time than the longest GSS series. Some questions have been repeated at 2-year intervals and others at 4-year intervals. Although these surveys have a political focus—providing data on political participation, support for the political system, and partisanship—many of the repeated questions provide data of interest to sociologists, social psychologists, and economists as well as to political scientists. For instance, the surveys have asked numerous questions on public policy issues such as those relating to health care, affirmative action, abortion, the military, and the economy.

The ANES project has its own Web site, from which the data can be downloaded, a guide giving selected data tabulations can be accessed, and weighting information is available. The weights provided are complicated and require careful study before cohort studies or other analyses covering a long time span are conducted. Some weights are available only for 1994 and later and can introduce incomparability with earlier data if they are used.

Several countries other than the United States have ongoing survey data collection projects similar to the American General Social Surveys and the American National Election Studies. Some of the resulting data are available from the Inter-university Consortium for Political and Social Research, but other data are available only from non-American data archives and Web sites.

A data collection project that has already gathered considerable data useful for cohort studies and that promises to be a major boon to cohort analysts in the future is the World Values Survey, a worldwide investigation of sociocultural and political change. Starting with the first wave of surveys in 1981–1984—with follow-up surveys in 1990–1993, 1995–1998, and 2000–2001—at least one survey with at least 1,000 respondents has been conducted in each of around 80 countries. About 20 countries have participated in all waves of the surveys, providing repeated cross-sectional data that can be used for cohort analysis. The surveys include a rich array of questions dealing with, among other topics, environmental issues, social norms, attitudes about marriage and family, gender issues, and political attitudes. Data from the first three waves are available from the Inter-university

Consortium for Political and Social Research, and the data set that includes data from the fourth wave became available for purchase as this edition of *Cohort Analysis* was nearing completion. The latest information about the project is available from the World Values Survey Web site.

In spite of their limitations, commercial opinion poll data are a valuable and underutilized source of data for cohort studies, often providing data gathered at short intervals rather than the 1- to 4-year intervals at which repeated academic surveys are conducted. It is hard to imagine a topic that has not been covered by questions on American opinion polls, and a good many questions have been repeated over a span of years or decades. The two main sources of poll data in the United States are the Roper Center for Public Opinion Research at the University of Connecticut and the Louis Harris Data Center operated by the Odum Institute for Research in Social Science at the University of North Carolina at Chapel Hill. The Roper Center is the oldest and largest poll data archive in the world, having most of the nonproprietary data collected by most of the major national polling organizations in the United States since 1935. Searches for data can be conducted online through the Roper Center Web site, but data sets are available for analysis only to persons affiliated with member institutions or who pay an acquisition fee. Most of the data collected by Louis Harris and Associates, one of the major American polling organizations, is available only from the Odum Institute, which has data from more than 1,200 Harris Polls conducted since 1958. Searches for data can be done through the Odum Institute Web site, and some data sets can be accessed online. Most other industrialized nations have at least one archive with an extensive poll data collection.

Unfortunately, age is coded only in broad categories in some earlier poll data sets, which makes them of very limited value for cohort studies. Cohort analysts searching for appropriate poll data should determine whether age (or date of birth) is coded in exact years before spending time and money acquiring data sets.

5. THE FUTURE

The future of cohort analysis is bright, primarily because the readily available data appropriate for this kind of research continue to accumulate, and there are already huge bodies of data that could be exploited to attain a better understanding of aging and social and cultural change. Undoubtedly, there will be methodological advances that will improve cohort analyses, although there will not be a magical statistical solution to the age-period-cohort conundrum. Cohort analyses can benefit from application of more complicated statistical techniques than the simple ones illustrated in this monograph, although those techniques cannot "solve" the identification problem. These techniques are mainly useful for other kinds of social research as well as for cohort studies, but new methods devised specifically for cohort analysis may prove useful.

Unfortunately, measurement advances in social research are likely to be of rather limited value to cohort analysis in the short run, because cohort studies must use older as well as recent data. Cohort analysts will continue to be subject to the criticism that the measures they use are not state of the art. However, cohort analysis can benefit from greater sensitivity to certain measurement issues, a past example being the growth in awareness of question order effects since the first edition of *Cohort Analysis* was published.

So far, almost all cohort analyses with survey data have used data from face-to-face interviews, but almost certainly, there will be more reliance eventually on data gathered by other means, such as by telephone. (Data from Internet surveys will not be useful for cohort studies in the foreseeable future.) Data gathered by means other than face-to-face interviews pose problems I do not discuss or only briefly touch upon in this monograph. For instance, age-related sampling incomparability may be different with telephone than with face-to-face surveys. I refer above to problems of comparability posed by using, in the same study, data gathered by different means of administering questionnaires, and there may be problems as least as serious involved in using telephone survey data gathered at different times. The increased use of answering machines and caller identification for screening telephone calls may be introducing new biases, and no one yet knows what the effects will be of the increase in exclusive reliance on cell phones by household members.

The logical issues facing cohort analysts will remain the same, but some of the technical problems will change, and may change rapidly, in the years ahead.

NOTES

1. Mason et al. fail to mention the possibility of a three-variable explanation for a linear pattern of variation in a dependent variable across age, period, and cohort.

2. One might wonder why such variables as percentage of cohort members whose parents divorced would be used in individual-level analyses if information on whether or not parents divorced is available for individuals. The reason is a hypothesized cohort contextual effect, by which children and adolescents whose parents did not divorce are affected by the presence of many other persons in their cohorts whose parents divorced. Of course, there may also be age-level contextual effects and period contextual effects.

3. Age effects are sometimes erroneously listed as sources of change. In fact, age effects are relevant to change only when there is a change in the age distribution of the population, which in turn occurs largely through cohort succession in any population appropriate for cohort analysis (see Chapter 4).

REFERENCES

ABRAMSON, P. R., & ENGLEHART, R. (1995). *Value change in global perspective.* Ann Arbor: University of Michigan Press.

ALWIN, D. F. (1991). Family of origin and cohort differences in verbal ability. *American Sociological Review, 56,* 625–638.

ALWIN, D. F., COHEN, R. L., & NEWCOMB, T. M. (1991). *Political attitudes over the lifespan: The Bennington women after fifty years.* Madison: University of Wisconsin Press.

BACHMAN, J. G., JOHNSTON, L. D., & O'MALLEY, P. M. (1976–2000). *Monitoring the future: Questionnaire responses from the nation's high school seniors.* Ann Arbor: University of Michigan Institute for Social Research. Twenty-five volumes (Ordering of compilers' names varies.)

BLALOCK, H. M., Jr. (1966). The identification problem and theory building. *American Sociological Review, 31,* 52–61.

BLALOCK, H. M., Jr. (1967). Status inconsistency, social mobility, status integration and structural effects. *American Sociological Review, 32,* 790–801.

CONVERSE, P. E. (1976). *The dynamics of party support: Cohort analyzing party identification.* Beverly Hills, CA: Sage.

DAVIS, J. A., SMITH, T. W., & MARSDEN, P. V. (2002). *General social surveys, 1972–2002* [machine readable data file]. Chicago: National Opinion Research Center.

FIREBAUGH, G. (1989). Methods for estimating cohort replacement effects. In C. C. Clogg (Ed.), *Sociological methodology 1989* (pp. 243–262). Oxford, UK: Basil Blackwell.

FIREBAUGH, G. (1990). Replacement effects, cohort and otherwise: Response to Rodgers. In C. C. Clogg (Ed.), *Sociological methodology 1990* (pp. 439–446). Oxford, UK: Basil Blackwell.

FIREBAUGH, G. (1992). Where does social change come from? Estimating the relative contributions of individual change and population turnover. *Population Research and Policy Review, 11,* 1–20.

FIREBAUGH, G. (1997). *Analyzing repeated surveys* (Sage University Paper series on Quantitative Applications in the Social Sciences, 07–115). Thousand Oaks, CA: Sage.

GLENN, N. D. (1974). Aging and conservatism. *Annals of the American Academy of Political and Social Science, 175,* 176–186.

GLENN, N. D. (1975). Trend studies with available survey data: Opportunities and pitfalls. In Social Science Research Council Center for Social Indicators, *Survey data for trend analysis: An index to repeated questions in U.S. national surveys held by the Roper Public Opinion Research Center* (pp. 6–48). Williamson, MA: Roper Center.

GLENN, N. D. (1976). Cohort analysts' futile quest: Statistical attempts to separate age, period, and cohort effects. *American Sociological Review, 41,* 900–904.

GLENN, N. D. (1977). *Cohort analysis* (Sage University Paper series on Quantitative Applications in the Social Sciences, 07–005). Beverly Hills, CA: Sage.

GLENN, N. D. (1980). Values, attitudes, and beliefs. In O. G. Brim, Jr., & J. Kagan (Eds.). *Constancy and change in human development* (pp. 596–640). Cambridge, MA: Harvard University Press.

GLENN, N. D. (1989). A caution about mechanical solutions to the identification problem in cohort analysis: Comment on Sasaki and Suzuki. *American Journal of Sociology, 95,* 754–761.

GLENN, N. D. (1994). Television watching, newspaper reading, and cohort differences in verbal ability. *Sociology of Education, 67,* 216–230.

GLENN, N. D. (1998). The course of marital success and failure in five American ten-year marriage cohorts. *Journal of Marriage and the Family, 60,* 569–576.

56

GLENN, N. D. (2003). Distinguishing age, period, and cohort effects. In J. T. Mortimer & M. J. Shanahan (Eds.), *Handbook of the life course* (pp. 465–476). New York: Kluwer Academic/Plenum.

HARDING, D. J., & JENCKS, C. (2003). Changing attitudes toward premarital sex: Cohort, period, and aging effects. *Public Opinion Quarterly, 67,* 211–226.

KIECOLT, K. J., & NATHAN, L. E. (1985). *Secondary analysis of survey data* (Sage University Paper series on Quantitative Applications in the Social Sciences, 07–053). Beverly Hills, CA: Sage.

MASON, K. O., MASON, W. M., WINSBOROUGH, H. H., & POOLE, K. W. (1973). Some methodological issues in the cohort analysis of archival data. *American Sociological Review, 38,* 242–258.

NAKAMURA, T. (1982). A Bayesian cohort model for standard cohort table analysis. *Proceedings of the Institute of Statistical Mathematics, 29,* 77-97 (in Japanese).

NAKAMURA, T. (1986). Bayesian cohort models for general cohort table analysis. *Annals of the Institute of Statistical Mathematics,* 38(Part B), 353-370.

O'BRIEN, R. M. (1989). Relative cohort size and age-specific crime rates: An age-period-relative-cohort-size model. *Criminology, 27,* 57–78.

O'BRIEN, R. M. (2000). Age-period-cohort-characteristic models. *Social Science Research, 29,* 123–139.

O'BRIEN, R. M., STOCKARD, J., & ISAACSON, L. (1999). The enduring effects of cohort characteristics on age-specific homicide rates: 1960–1995. *American Journal of Sociology, 104,* 1061–1095.

PALMORE, E., & JEFFERS, F. C. (1971). *Prediction of life span.* Lexington, MA: D. C. Heath.

RODGERS, W. (1990). Interpreting the components of time trends. In C. C. Clogg (Ed.), *Sociological methodology 1990* (pp. 421–438). Oxford, UK: Basil Blackwell.

SASAKI, M., & SUZUKI, T. (1987). Changes in religious commitment in the United States, Holland, and Japan. *American Journal of Sociology, 92,* 1055–1076.

SASAKI, M., & SUZUKI, T. (1989). A caution about the data to be used for cohort analysis: Reply to Glenn. *American Journal of Sociology, 95,* 761–765.

U.S. CENSUS BUREAU. (1969). *Statistical abstract of the United States: 1969.* Washington, DC: U.S. Government Printing Office.

U.S. CENSUS BUREAU. (1979). *Statistical abstract of the United States: 1979.* Washington, DC: U.S. Government Printing Office.

U.S. CENSUS BUREAU. (1990). *Statistical abstract of the United States: 1990.* Washington, DC: U.S. Government Printing Office.

U.S. CENSUS BUREAU. (1999). *Statistical abstract of the United States: 1999.* Washington, DC: U.S. Government Printing Office.

INDEX

58

ABOUT THE AUTHOR

Norval D. Glenn is the Ashbel Smith Professor in Sociology and Stiles Professor in American Studies at the University of Texas at Austin. His main research interests relate to aging and the life course, and family relations in modern societies. He is a former editor of *Contemporary Sociology* and the *Journal of Family Issues*, and he has served on the editorial boards of such journals as the *American Sociological Review, Public Opinion Quarterly, Journal of Marriage and Family, Demography,* and *Social Science Research.*

His recent publications deal with such topics as the dissemination of social science findings to policy makers and the general public, changes in the institutional mechanisms of mate selection, and the relationship of age at marriage to marital success.